Astronomer's Pocket Fiel[illegible]

MW00655744

For further volumes:
http://www.springer.com/series/7814

Jamey L. Jenkins

Observing the Sun

A Pocket Field Guide

Springer

Jamey L. Jenkins
Homer, IL, USA

ISBN 978-1-4614-8014-3 ISBN 978-1-4614-8015-0 (eBook)
DOI 10.1007/978-1-4614-8015-0
Springer New York Heidelberg Dordrecht London

Library of Congress Control Number: 2013942651

Cover illustration: ESA/AOES

Springer is part of Springer Science+Business Media (www.springer.com)

Preface

There are many books on today's market that cover what is undoubtedly the most energetic body in our solar system, the Sun. Many of these titles are a great read, and answer a number of the questions asked by observers regarding this truly dynamic subject. These are books which make excellent additions to any observer's library as they target the armchair astronomer, one who enjoys reading on an overcast day.

That said, this book has a different resolve. The aim here is to provide you with a source that can be taken to the observing field, and when needed referred to at the telescope. It will fit conveniently in a pocket for easy access, or in a box of accessories, hence the compact size. When I was first approached with the idea of producing a field guide for the Sun, a light bulb immediately lit over my head. The concept of a pocket-sized guide for the solar observer was admirable. As far as I knew, this had never effectively been done before.

Because the Sun is an object in constant flux, astronomers rarely know for certain what awaits them at the telescope. All solar features are transient, sometimes fleeting, and the life cycles of most tend to be somewhat elaborate; in those cases, classification schemes become intricate. Take for instance sunspots, which begin as pores in the photosphere. Most decay but others sometimes continue development, growing in girth, strength, and numbers. These specimens in turn sprout penumbra and conceivably can become monstrous complex blemishes on the solar face. Spots develop into a variety of types.

Committing the stages of a life cycle of such features to memory can be a daunting challenge. Other times, an observer asks, "What type of prominence am I viewing, and why is it so bright?" This book is designed to help answer such questions, and many more. The amateur solar astronomer is in need of a comprehensive guide to identify features, one that is intended for use in the field.

As a consequence, I have written *Observing the Sun: A Pocket Field Guide*.

The book is divided into two distinct components to facilitate its use. Part One focuses closely on the Sun and its features, while observational techniques are discussed in Part Two. You will not have to search exhaustively for information about white light features, chromospheric activities, or the solar corona. That information is organized into separate sections within appropriate pages. Each piece, examining the photosphere, lower and upper chromosphere, and the corona, opens with a keyed whole disc image pointing out significant phenomena. Then the following pages contain descriptive text, many times accompanied by close-up illustrations depicting the discussed feature.

To effectively use this field guide at the telescope, begin by reading the material ahead of time, and become acquainted with solar features before ever placing your eye to the telescope. You should possess a general idea of what to expect when you observe the Sun, or even the most obvious structures will escape you. This is called *developing your observing eye*.

Next, decide what part of the Sun to observe. By part I don't mean location on the disc, but rather what layer. Solar observing is viewing the Sun's atmosphere. The atmosphere of the Sun is divided into several unique layers that are best seen by an amateur solar astronomer using three different wavelengths of light. The lowest atmospheric layer is the photosphere often erroneously, but acceptably, called the surface. The photosphere only appears surface-like because of its opacity. The other layers above the photosphere include the chromosphere, which will be divided into its lower and upper regions, and lastly the corona. The heart of this book has four sections each dedicated to one of the layers. These sections are presented progressively as they exist on the Sun from the photosphere outward.

Discerning the photosphere is done by observing in the solar continuum, commonly called white light. We see the region just above the photosphere (lower chromosphere) in the light of calcium, using Ca-K appliances and the upper regions of the chromosphere best with H-alpha viewing appliances. The fourth layer, the solar corona, is visible to an amateur astronomer only during a total solar eclipse, once more best in white light.

Depending on the layer being observed, identify the corresponding section of the field guide to be referenced. Features within their respective section are organized in a pattern of observability and development. For instance, the photosphere section begins with simple naked-eye observations and progresses to granulation, faculae, pores, and then the most recognizable feature, sunspots. A similar progression is found in the other sections.

The illustrations accompanying the dialogue will prove invaluable for identification at the telescope. An excellent example is found with the Roel image attending the pores and voids commentary. A pore does not significantly appear different from dark granular material. With an illustration handy, no doubt remains as to which is a pore and which is not.

This book is more than an illustrated guide to solar phenomena; within Part Two you'll find information on how to observe the Sun safely. With today's modern tools and a common sense approach, solar observing is a 100 % safe activity. Instruments for observing the Sun as well as explanations of solar directions, rotations, and numbering systems are covered. Knowing what to do with an observation is as important as knowing what you are seeing; therefore the second Part also guides the reader in simple steps for recording solar observations.

The book concludes with a thorough Appendix section composed of current organization websites, a glossary, blank recording forms, and Stonyhurst templates that can be photocopied for your record-keeping. An ephemeris of daily solar statistics is generously included containing the elements required for reduction of daily observations.

The Sun in my opinion *is* the most interesting object we have to study in the heavens. Some time ago, I conducted a not-so-scientific survey of online friends to discover what they found fascinating about observing the Sun. Of course many responded relating to the sheer dynamics of the Sun, or the beauty and magnitude of the events happening there. Someone else stated humorously that, "the Sun is easy to find in the sky." And yet others declared, "It is simply fun to share views with the public."

Solar observing is indeed fun, but a word of CAUTION is always in order to prevent the novice from immediately rushing out and directing a telescope Sunward. The Sun emits huge quantities of heat, light, and radiation, which **must** be respected at all times. The greatest danger to the Earth-bound observer is in the brightness and IR/UV light of the Sun. These non visible wavelengths must be blocked and the intensity of the solar illumination reduced significantly for safe studies to be conducted. Without these precautions, **blindness of the observer will result.** This topic is discussed in following pages. Regardless, the author and publisher cannot be held responsible for the careless action of any solar observer disregarding safety procedures. The rule of thumb regarding solar observing is to always err on the side of safety. Do that and you'll enjoy many years of viewing nature's most magnificent spectacle.

Homer, IL, USA Jamey L. Jenkins

Acknowledgments

This book owes its creation to the efforts of a number of individuals to whom I express my sincerest appreciation. First comes to mind, my dear friend from the past, the late Howard F. Zeh for his never failing encouragement and mentoring, without which my fledgling interest in amateur astronomy may never have bloomed into what has now become a lifelong hobby. Next would be Richard Hill of the Lunar and Planetary Laboratory, University of Arizona, who has provided the inspiration that led me to this project. Thanks for your support Rik!

Gratitude of an immeasurable amount goes to the army of observers stationed around the world who generously supplied their expert imagery of solar features. Without each photographer's effort this project would have been but a collection of words and thoughts. Their pictures tell the story so much plainer than I ever could articulate. Contributing solar photographers to this project include: Art Whipple, Eric Roel, Alexandra Hart, Christian Viladrich, Fabio Acquarone, Jim Ferreira, Thomas Ashcraft, Gordon Garcia, Steve Rismiller, and Dr. Neal Hurlburt of the Lockhead Martin Solar and Astrophysics Laboratory for the TRACE image.

Several contributors of graphic material or photos of their specialty equipment also need recognition. These include William Waller, Raymond Rienks, Greg Piepol, Robert Hess, and Dr. David Hathaway of NASA.

A special thank you to Brad Timerson of the Association of Lunar and Planetary Observers Solar Section for his computation of the Solar Ephemeris found in the Appendix of this book. Brad has *always* cheerfully come through with this vital information the several times I've requested data.

My family deserves the greatest indebtedness of all for their unconditional support during the many hours spent behind the keyboard focused on this venture. And to my wife Mary, thank you for believing in my desire to share solar astronomy with an interested readership.

And lastly, to those who have obtained a copy of this book. Without you the dissemination of information would have been futile and served no purpose. Enjoy astronomy, and in particular solar astronomy, where from your own backyard you may find delight in an intimate view of our nearest star, the Sun.

Homer, IL, USA Jamey L. Jenkins

About the Author

This is the second book written by Jamey Jenkins about the Sun. Springer published the first, *The Sun and How to Observe It*, in 2009 as a broad look at solar observing. That book took a wide-ranging approach, explaining to the amateur astronomer the how and why of studying the nearest star. This venture, *Observing the Sun: A Pocket Field Guide* is meant for decided use at the telescope with a specific focus on the Sun's abundant features.

A product of the space age during the heyday of the 1960s' Gemini and Apollo space programs, Jenkins' experimentation with a succession of increasingly larger telescopes led to developing several pen pal friendships and an invite to write for Dave Eicher's fledgling amateur journal, *Deep Sky Monthly*. As time passed, work and family responsibilities began assuming a priority and evening observing soon succumbed to the daytime study of sunspots and related solar phenomena.

Jenkins has contributed to the Sunspot Program of the American Association of Variable Star Observers (AAVSO) and is an active member of the Association of Lunar and Planetary Observers (ALPO) Solar Section. He's also served as Assistant Section Coordinator of that group for a number of years. Jenkins firmly believes that it is vital for one wishing to develop in this hobby to have a connection with other observers; a mentor is invaluable when it comes to understanding technical aspects.

These days, Jenkins photographs sunspots, watches calcium clouds, and studies prominence activity from his backyard observatory with a substantial 125 mm f/18 refractor. A significant development from the tiny Galilean lunar telescope of his past, this home-assembled telescope shows the Sun's unique character, as a seething, boiling cauldron of gas, and indeed the master of the solar system.

Contents

PART I

THE SUN

CHAPTER 1

THE SUN IS A STAR

You are no doubt familiar with the evening sky, and the wonders it presents as twilight deepens to darkness. A typical night may find the Moon displaying its ghostly earthshine along with the major planets glowing brightly during their wander through the zodiac. Occasionally, meteors flash and hazy comets or star-like asteroids slip quietly about the solar system. Providing a backdrop for these menageries you also find an infinite quantity of seemingly fixed points of light, the stars.

Through the largest telescopes, stars appear as singular points of light. They have unique colors and characteristics, but all are so distant we cannot discern their faces. Many are paired with other stars doing an orbital dance in the sky, so-called double and multiple star systems. Stars are not always steady in brightness either, some pulsate with the regularity of a pocket watch and others explode unexpectedly with the power from many millions of nuclear bombs. Nova and supernova catch us all by surprise. Recent discoveries have placed gas, dust clouds, and planets around some of these far away Suns. Yes, Suns! Stars are but huge balls of glowing gas like our Sun. Each star has unique properties, but unfortunately all are too far away to be seen clearly.

Our Sun, however is a star that we can see up-close and personal. Being as near as we are, it provides a laboratory right in our backyard for discovering how stars similar to it function. In this chapter we will discuss how it is believed the Sun came to be, what happens inside the Sun to make the light and energy we require to exist, and lastly how the Sun effects the nearby space environment of the Earth.

Jamey L. Jenkins, *Observing the Sun: A Pocket Field Guide*,
Astronomer's Pocket Field Guide, DOI 10.1007/978-1-4614-8015-0_1,

Solar Origin

The Sun found its origin in the humble beginnings of a vast cloud of dust and gas called the solar nebula. This nebula or cloud was a collection of the numerous elements found in today's solar system including hydrogen, helium, carbon, nitrogen, oxygen, neon, magnesium, silicon, sulfur, and iron. Traces of the elements nickel, calcium, argon, aluminum, and sodium, as well as gold were also present.

Hydrogen and helium make up nearly 98 % of the mass of the Sun; the other elements originated in the interiors of early stars and were dispersed throughout our galaxy, the Milky Way, via exploding supernova. It's amazing to contemplate that all we see around us and even our bodies were at one time inside distant ancient stars.

Astronomers believe that dusty particles found within the frigid temperatures of the solar nebula were coated with an elemental icy compound. Over time gravity would cause a tendency for these icy particles to migrate toward the center of the nebula, creating a region of greater density and pressure termed, the protosun. Inside the protosun temperature began to rise because of the crowding together of atoms, which haphazardly bounced off of one another. This process, the turning of gravity's energy to heat, is called Helmholtz contraction. An angular momentum or rotation of the solar nebula had to be present or all matter would eventually have been drawn into the emerging protosun. It could have been the shockwave of a nearby supernova explosion or perhaps just an inherent characteristic of the matter that caused the rotation of the nebula. Nevertheless, with passing time the growing pressure and increasing temperature of the contracting gas and particles would cause the infant Sun to "ignite" and begin to glow.

Gravity and the Helmholtz contraction were responsible for the protosun; but could this process sustain a star's appetite for energy given the known age of our Sun? No, if contraction were indeed the sole source of the Sun's energy, as was initially thought in the nineteenth century, the Sun should have exhausted itself after only millions of years. However, we know the Sun has been around for several billion years. The ultimate source of the Sun's energy could not be gravity and Helmholtz contraction.

Table 1.1 The Kelvin scale of temperature

–	Kelvin	Celsius	Fahrenheit
Absolute zero	0 K	−273 °C	−459 °F
Freezing point of water	273 K	0 °C	32 °F
Center of the Sun	15,000,000 K	14,999,727 °C	26,999,540 °F
Surface of the Sun	5,800 K	5,527 °C	9,980 °F

Near the beginning of the twentieth century, physicists discovered that under conditions of tremendous temperature and pressure, hydrogen atoms fuse together to form helium. In the course of this process a small portion of an atom's mass is released as energy. This freeing of energy explains how the Sun is energized deep in its bowels, in a region designated the core. A nuclear process is the source of the Sun's power.

It is estimated that 20 % of the solar radius is engulfed by the core. The pressure there is nearly 340 billion times the atmospheric pressure at sea level on Earth and the temperature in this region exceeds 15 million Kelvins. To give a perspective, consider that water boils at about 373 K! (Table 1.1).

Conditions are unimaginable in the core with gas having a density much greater than lead and a temperature so extreme that electrons have been stripped from their atoms. This stripping process is dubbed ionization. An atom having one or more electrons missing is an ion. The Sun's atoms are totally into a soupy mixture of ions and electrons labeled plasma.

Physicists tell us that a star having a mass equal to or less than the Sun goes through a process of converting hydrogen (98 % of the Sun's mass) to helium via the proton-proton cycle (PPC). In simplest terms, during this conversion four hydrogen atoms are transformed to a single helium atom with a portion of the mass released as high-energy photons. The PPC powers the Sun's internal furnace.

The proton-proton cycle results in the conversion of millions of tons of hydrogen to helium in the Sun every second. Some day after all the hydrogen is gone; the outer layers of the Sun will be blown away. As seen from distant space our solar system will be what we today know as a planetary nebula. No need for immediate concern. With the hydrogen to helium conversion already underway for nearly 4.6 billion years, it is estimated that approximately another 5 billion years of hydrogen is available.

Structure of the Sun

The Sun is powered through a nuclear fusion process deep within its core. Thankfully, it doesn't react like a gigantic hydrogen bomb, bursting and destroying everything within the immediate solar system. Without the outward pressure from the compressed gas, gravity would collapse the external layers of the Sun onto the core. This arrangement, hydrostatic equilibrium, keeps everything in a relative constant state. Another important component to a steady functioning Sun is the uniform conversion of hydrogen to helium. The so-called hydrogen burning doesn't stop, restart, or vary in its regularity. The continuous fuel-in, energy-out principle we know as thermal equilibrium. Without these two principles, hydrostatic equilibrium and thermal equilibrium, the Sun could not exist in its present uniform state.

Energy transmission from the core to the Sun's outer layers and beyond is comprehendible when we begin with an awareness of how the Sun is structured. Consider briefly an onion as obtained from the market. Peel away the outer flaky shell and you come to the skin. With a knife or thumbnail peel away the uppermost layer of the onion and you find yet another layer, and another, and so on. The Sun is a body, layered a bit like the onion. Beginning at the center and working our way out, the first layers of the Sun are: the core, the radiative zone, and the convection zone. Next we find the photosphere, the chromosphere, the transition zone, and lastly extending far into space, the corona.

Radiative Zone

Earlier we determined that the core is the nuclear powerhouse of the Sun. Above it and extending out to a point some 70 % of the solar radius is the radiative zone. Here the temperature has cooled to approximately 8,000,000 K and is less dense as compared to gas in the core. High-energy gamma ray photons released by the fusion process begin their long journey to the surface from the core through the radiative zone. The natural tendency of a photon is to depart a hotter region and seek a cooler region. However, gases present in the radiative zone absorb the photons, reradiating them in directions not always conducive to movement toward

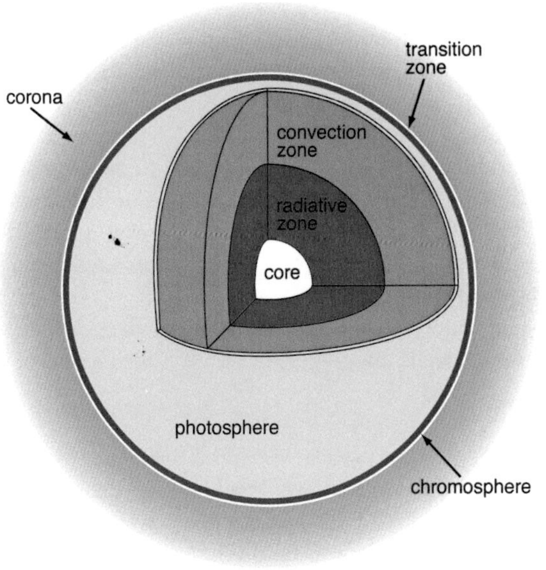

Fig. 1.1 Layers of the Sun from the inside out are the core (*white*), radiative zone (*red*), convection zone (*orange*), the photosphere (*yellow*), chromosphere (*magenta*), transition zone (*white*), and reaching into space becoming one with the solar wind, the corona (*green*)

the surface. A photon can be absorbed and re-emitted 10^{25} times before departing this region. Consequently, it may take as much as 100,000 years or more for a photon to break loose from the radiative zone (Fig. 1.1).

Convection Zone

The remaining 30 % of the Sun's radius located directly above the radiative zone is the convection zone. Energy is transported outward by deep rising currents in the convection zone to the Sun's surface, the photosphere. Understand the Sun has no true surface; it is a gaseous body of decreasing density from the center outward. The photosphere simply defines the assumed edge of the Sun only because it is as deep as we can look into the opaque solar interior.

Energy transfer in the convection zone is analogous to a few everyday activities. For example, think of a boiling pot of water. When heat is generated at the bottom of a pot, tiny pockets of thermal energy collect and rise through the water producing the bubbling effect we call boiling. This is convection in action. A very similar occurrence is detected on the Sun. The photons produced in the core and passed through the radiative zone create pockets of hot gas (plasma) in the convection zone that rise to the solar surface. On the Sun, each 1,000 km diameter pocket (granule) makes its way to the top of the convection zone (photosphere) at about 1,500 km/h. Upon arrival in the photosphere a releasing of energy takes place followed by the granular material cooling as it flows back to the solar interior along the granule's outer walls. The cooler, darker walls define a granule's unique kernel shape.

Another movement of the plasma within the convection zone, other than vertical convection, has been shown. This movement, the meridional flow, causes plasma to travel in a loop from the equator to the polar region and back again. Because sunspots are conjectured to have a magnetic tie to the lower region of the convection zone, the gradual migration pattern of the meridional flow coincides with the well known 11-year solar cycle.

Interestingly, while the core and radiative zone rotate like a solid body possessing a period of about 27 days, the convection zone acts as though it were a liquid having an equatorial rotation period of about 25 days and a polar interval of nearly 36 days. Shearing (drag or friction) between the solid acting radiative zone and the fluid-like convection zone helps to create the Sun's magnetic field. The region where these zones meet is the tachocline.

Plasma's nature as it moves is to create a magnetic field. The lines of the Sun's magnetic field are parallel to its axis of rotation, from north to south. Differential rotation wraps the internal magnetic field around and around the Sun as the magnetic field lines are dragged along with the charged particles of plasma. Over time the vertical motion of the plasma in the convection zone causes a kinking or tangling of the elongated magnetic field lines. The tangled lines develop regions of increased magnetic strength in the convection zone known as magnetic flux tubes. Although it is still unclear as to what causes the flux tubes to rise to the solar surface, most solar astronomers attribute it to buoyancy. Small flux tubes poke through

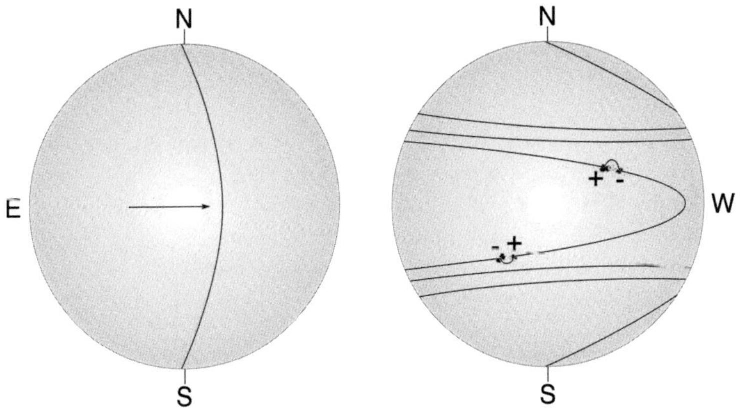

Fig. 1.2 Babcock's sunspot theory teaches that magnetic field lines running from the north to south polar regions are dragged around and around the Sun until the lines tangle, strengthening themselves and poking through the photosphere as flux tubes. Convection is stifled thereby forming sunspots

the solar surface at bright points, forming chains known as filigree. Larger flux tubes are darker and produce pores and sunspots (Fig. 1.2).

Photosphere

The photosphere (meaning sphere of light) is the beginning of the solar atmosphere, and the lowest level we can visually see into the Sun. You will find the surface temperature a sweltering 5,800 K. Photons exit this layer and literally flood our world with the heat and light necessary to sustain life.

The photosphere is discernable to a depth of about 500 km. The underlying gas has become so dense and opaque that it is impossible to detect what is happening below with conventional telescopes. Astrophysicists measure wave oscillations of the photosphere with a science known as helioseismology. These oscillations are used to study the temperature, motion, and composition of the Sun below its surface to regions near the core. Such studies are beyond our scope, but as amateur astronomers we can observe many photospheric activities including sunspots, faculae, and granulation.

Solar features experience a cyclic pattern of waxing and waning in their frequency called, the 11-year solar cycle. First noted in the early nineteenth century by S. H. Schwabe, a German pharmacist searching for an inter-Mercurial planet, the newly discovered cycle took everyone by surprise. Schwabe originally announced the periodicity as being 10 years, but this was refined after further observations to an average of about 11.1 years, and the cycle seems to vary between 8 and 14 years.

For the amateur solar observer this is where the fun begins. Magnetic fields generated due to differential rotation poke through the photosphere, stifling the bubbling convection process, and creating cooler dark regions we refer to as sunspots. Lending themselves to several classification schemes because of their myriad variations in appearance, sunspots provide hours of relaxed study for the solar observer. A statistical examination called sunspot counting is popular with many observers. Observations of sunspot motions and their changing appearance (morphology) can be documented with drawings or by using modern imaging techniques.

Faculae are cloud-like patches often seen near a sunspot group, usually as a precursor to an emerging sunspot. While a sunspot is a large, dark-appearing region of magnetic strength; a facula is a magnetically weak, granular sized region. Faculae are better seen near the Sun's edge as they stand out against limb darkening. Limb darkening itself gives the Sun a definite three-dimensional appearance causing the Sun to appear brighter nearer the center of the disc. It is an effect created by the cumulative opacity of sub-surface gas viewed toward the Sun's limb.

Chromosphere

Above the photosphere is the chromosphere, with a thickness approaching 2,000 km and a variation of temperature from 6,000 to 20,000 K. Chromospheric gas is rarefied or less dense than that found below, and has a certain pinkish-red hue. During a total eclipse of the Sun the Moon temporarily blocks the photosphere and the chromosphere becomes visible to our eyes as the magenta colored ring hugging the solar limb.

The chromosphere is reddish because of its emission strength at 656.3 nm, which is the wavelength of the Hydrogen-alpha line in the solar spectrum. In other words, Hydrogen-alpha is the point in the spectrum that shows ionized hydrogen the best. The chromosphere is not easily seen because it is so brightly outshone by light from the photosphere. Astronomers using special instruments that allow the light of Hydrogen-alpha to pass and blocking the other wavelengths are able to study features of the chromosphere without the aid of an eclipse.

Solar activity in the chromosphere takes many forms. Spicules are easily visible at the limb as small jets of gas shooting out of the Sun. They have also been described as spikes or sometimes as resembling blades of grass. On the disc of the Sun, spicules are called mottles, and at other times fibrils. Spicules assume patterns that have been likened to bushes, chains, and starry-shaped rosettes. The cause of spicules has not been well defined, but they do have a typical lifetime of about 15 min and may reach altitudes of 3,000–10,000 km.

Another form of chromospheric activity, the prominence, should not be confused with a spicule, as a prominence is a cloud or stream-like projection of hot gas (10,000 K) found above the solar limb. Prominences exist because hot ionized gas clings to the strong magnetic fields of the Sun; consequently the gas traces out lines and loops where magnetic fields are present. The fields originate below the photosphere and with sufficient strength extends into the upper layers of the solar atmosphere. If seen silhouetted against the darker background sky, prominences appear bright. Viewed against the backdrop of the brilliant Sun, a prominence looks dark and is then known as a filament.

As varied in appearance as sunspots, prominences are also subject to classification schemes. In the simplest scheme they are divided into two basic classes, *quiescent* and *eruptive*. Eruptive prominences are subject to rapid motions. Sometimes known as active, an eruptive prominence has a short lifetime, perhaps on the order of a few hours, and is often ejected violently from the Sun. Quiescent prominences on the other hand have much longer lives, often many days, and take shapes as hills or trees at the solar limb. Quiescent prominences are slow to change appearance, hence they are deemed quiet.

Flares are the sudden release of built up stress in the magnetic field of an active region. A solar flare many times appears as the intense brightening of a bright patchy area in the chromosphere dubbed a plage. The initial phase of brightening can be rather quick, from a few minutes to an hour. Then a gradual decline in brightness occurs as the flare loses strength. The energy output of a solar flare can be truly astronomical. A fair comparison would be for a large flare to match the total output of the Sun for a period of several seconds. Compress this energy into the relatively small flaring area on the Sun and the magnitude of the event becomes obvious.

Transition Zone

There exists an irregularly shaped layer several 100 km thick that separates the chromosphere from the above corona. This region is known as the Transition Zone. The temperature of the gas here does an abrupt leap from the chromospheric temperature (20,000 K) to that of the immediate corona (1,000,000 K). This region's visibility is inaccessible to the amateur astronomer since its emission largely contains the elements carbon, oxygen, and silicon in wavelengths visible only to satellites from outer space (Fig. 1.3).

Corona

Beyond the chromosphere and transition zone is the pearly white outer atmosphere of the Sun, the corona or crown. Compared to the photosphere the light of the corona is extremely weak, becoming even weaker with distance. Visible only during a total eclipse of the Sun or through a special instrument called a coronagraph, the best views of the corona are obtained from spacecraft, which orbit above the scattering effects of dust, refraction, and water vapor found in Earth's atmosphere.

The shape of the corona varies with the 11-year solar cycle. During sunspot minimum the corona is usually found restricted in the Polar Regions. However, near sunspot maximum we discover the corona "blooming" and encompassing the entire solar disc.

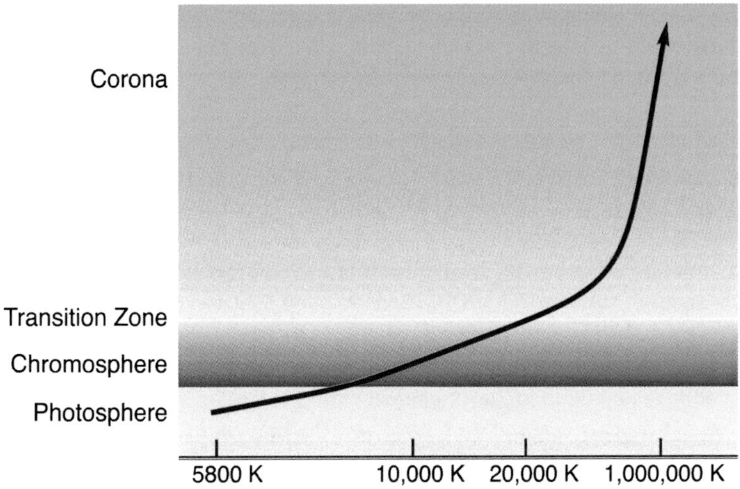

Fig. 1.3 Temperature increases dramatically as the Transition Zone is crossed, reaching in the upper corona several million Kelvins

Coronal temperature rapidly exceeds 500,000 K and at some distance from the photosphere may approach over 2,000,000 K. The heating mechanism is unknown and remains one of the big questions to be answered by solar astronomers.

As the corona stretches farther and farther from the Sun, it gradually becomes one with the charged particles streaming away designated the solar wind. In a way, we reside within the confines of the Sun, living with the never-ending stream of its particles.

Spaceweather and Earth's Magnetosphere

Spaceweather is a relatively new term used to describe the near Earth space environment as affected by energy and particles from the Sun. Understanding spaceweather is vital to our everyday lives because of the adverse effect solar energies can have on health issues, communication, and our power grid structure.

Our planet behaves as though it were a giant bar magnet with lines of magnetism emerging from the north and south poles, arching tens of thousands of kilometers into space and returning to the opposite pole. The iron core of our planet plus the dynamo effect creates the Earth's magnetic field. Normally, we think of magnets as having attracting power, but repulsion is also a characteristic of magnetism. Objects like the solar wind that have a charge and are moving will be repelled or pushed away by a magnetic field. There is a pit or hollow in the solar wind created by the Earth's magnetic field known as the magnetosphere. As the solar wind encounters the magnetosphere, the outside border called the bow shock immediately buffers it. The inner borderline is referred to as the magnetopause; here the bulk of the solar wind is repelled around the Earth's magnetic field, in a chaotic region dubbed the magnetosheath. See Fig. 1.4 for a graphic depiction of how Earth's magnetic field affects the solar wind.

Solar Wind and CMEs

The Sun expels a stream of charged particles and bits of matter, the solar wind, which flows throughout the solar system with a velocity nearing 400 km/s, pushing against Earth's magnetosphere and flattening its Sun-facing side. As these charged particles encounter the magnetosphere they are repelled, saving the Earth and its inhabitants from certain annihilation. However, during especially energetic events such as a solar flare, particles sometimes get caught up in our protective shield, disrupting the magnetosphere and adversely affecting the local spaceweather.

As the Sun pumps large amounts of X-ray, Ultra-violet, and gamma ray radiation into space, Earth's aurorae (Borealis in the north and Australis in the south) become active. Particles trapped in the field lines of the magnetosphere produce electric currents that travel to Earth's magnetic poles. Gases (oxygen and nitrogen) found in the upper 80–300 km of atmosphere are excited and begin to glow. Excited oxygen atoms glow red and green, nitrogen glows red or blue. While the aurora are a beautiful result of the Sun's activity, other rather negative effects may include: the destruction of delicate electronics in orbiting satellites, upper atmospheric changes that place unanticipated drag on satellites, electronic (TV, radio, and telephone) transmission disruptions, and the forming of holes in Earth's ozone layer.

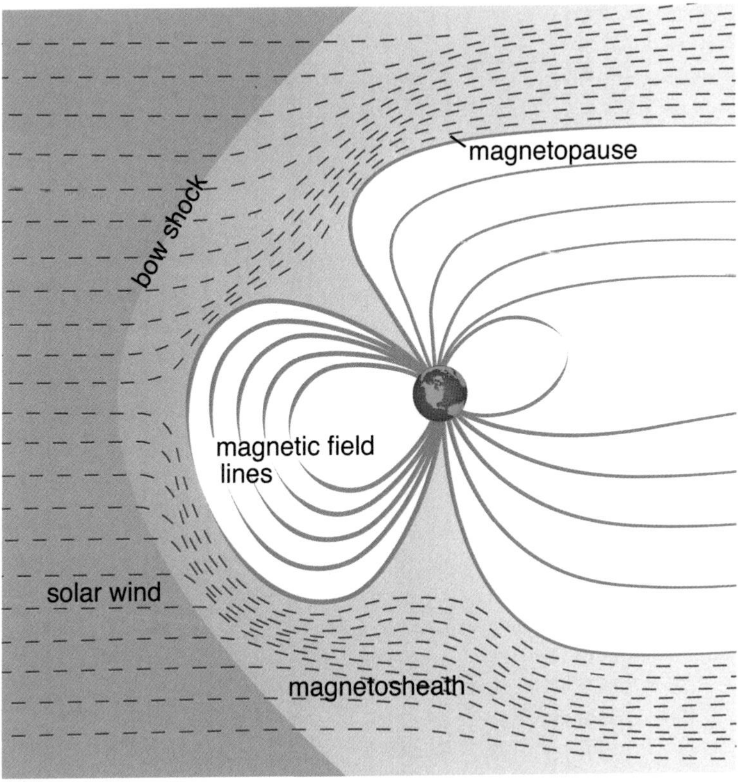

Fig. 1.4 The magnetic field of the Earth creates the magnetosphere. Comprised of a bow shock, magnetosheath, and magnetopause, the solar wind is repelled around the Earth (Illustration courtesy of William Waller)

Created in the upper atmosphere, ozone absorbs UV light protecting us from increased rates of skin cancers, cataracts, and other diseases.

Sometimes an event called a coronal mass ejection, or CME, takes place. A CME is a huge bubble or balloon shaped pocket of gas ejected from the Sun. Not discovered until the early 1970s when orbiting spacecraft began systematic studies of the Sun, a CME often represents the loss of several billion tons of solar matter. During the decline of the solar cycle the frequency of CMEs is about one per week. However, solar maximum brings about two or three CMEs daily! Fortunately the majority are not directed

Table 1.2 Facts regarding the makeup of the Sun (Data courtesy of NASA)

Diameter: 1,391,980 km (109 Earth diameters)		
Mass: 1,989,100 × 10^{24} kg (333,000 Earths)		
Volume: 1,412,000 × 10^{12} km^3 (1,304,000 Earths)		
Visual magnitude: −26.74		
Absolute magnitude: +4.83		
Spectral type: G2V		
Distance from Earth	Minimum	147,100,000 km
	Mean	149,600,000 km
	Maximum	152,100,000 km
Apparent diameter	Minimum	31.4 min of arc
	At 1 A.U.	31.9 min of arc
	Maximum	32.5 min of arc
Central pressure: 2.477 × 10^{11} bar		
Central temperature: 1.571 × 10^{7} K		
Central density: 1.622 × 10^{5} kg/m^3		
Central composition: 35 % H, 63 % He, 2 % C, N, O…		
Photospheric pressure (top): 0.868 mb		
Photospheric temperature (top): 4,400 K		
Photospheric effective temperature: 5,778 K		
Photospheric temperature (bottom) 6,600 K		
Photospheric composition: 70 % H, 28 % He, 2 % C, N, O…		
Sidereal rotation period: 25.38 days		
Synodic rotation period: 27.27 days		
Age: 4.57 × 10^{9} years		

toward the Earth, but are visible in profile at the solar limb or are invisible as they develop on the far side of the Sun.

Solar flares often trigger CMEs. Active regions home to sunspots harbor the precipitating flares, but occasionally a CME does result where no flaring is happening. Viewed head-on, a large CME directed toward the Earth is called a halo event, an apt name since the expanding shell of gas appears to envelop the Sun in a beautiful "halo" (Table 1.2).

CHAPTER 2

IDENTIFYING SOLAR FEATURES

For the serious amateur astronomer, observing the Sun is all about measuring, counting, or recording solar features that evolve on a continuous basis. Observations tend to split into two distinct areas of study called statistical and morphological. Statistical efforts are tabulated by an observer following various solar metrics like the daily sunspot number, prominence frequency, or the noting of sunspot group classifications. Morphology is documenting the changing appearance of a feature. Solar morphology is inclusive of photos/drawings that depict features and their positions on the Sun or perhaps a series of images showing the eruption of a prominence, even an individual snapshot of an active region.

To conduct these studies adequately it is necessary to be familiar with solar features. Features of the Sun are some of the most dynamic found in the solar system, each having certain defining features. Beginning first with the photosphere and moving outward through the chromosphere to the corona, let's see what the Sun has to offer.

Photosphere

The lowest layer of the solar atmosphere is the photosphere. It is observable in the solar continuum, more commonly called *white light*, via direct (objective filter/Herschel wedge) or indirect (solar projection) methods. Interestingly, due to opacity, the photosphere appears to give a solid surface to the gaseous globe of the Sun. Evidence of convection from the region below is provided by the granules best visible near the center of the solar disc. Magnetic fields make their presence known by inhibiting convection, thereby developing faculae and dark sunspots (Fig. 2.1).

Jamey L. Jenkins, *Observing the Sun: A Pocket Field Guide*,
Astronomer's Pocket Field Guide, DOI 10.1007/978-1-4614-8015-0_2,

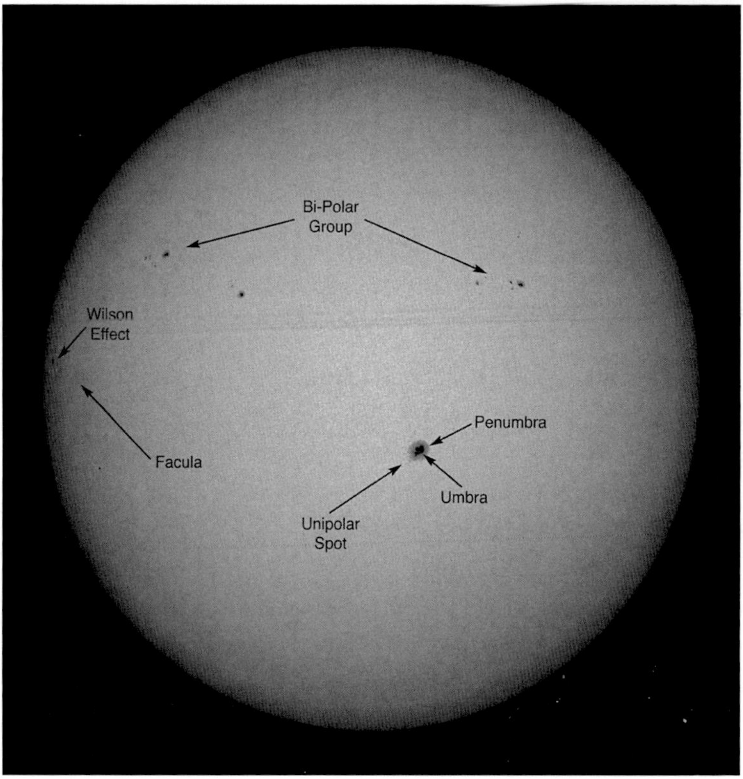

Fig. 2.1 Solar photosphere in white light from 6 April 2013. The tiny symmetrical spot on the east (*left limb*) is exhibiting the Wilson effect (Illustration courtesy of Alexandra Hart)

Naked Eye Sunspots

The existence of blemishes on the Sun has been known for several thousand years thanks to the interpretations of ancient Chinese astronomers. Observations were made by looking unwittingly at the Sun with the naked eye during occurrences of thin clouds, fog, or when atmospheric extinction happened to be heavy at sunrise or sunset. Consequently, the Chinese were able to achieve these first sunspot sightings. Of course today we know their techniques to be a very risky and dangerous business. We must not attempt to duplicate their observations without the proper filtration protecting us from the Sun's harmful effects.

To view sunspots with the naked eye safely, a white light objective filter of visual density 5.0 is used. A circular filter works famously if it is large enough to envelop your face, for which 125 mm diameter is usually sufficient. Cup your hands around the edge of the mounting cell to limit ambient light from reaching your eyes, hold it before your face and take a deep look at the Sun. The visibility of a borderline spot can be improved by rocking your head from side-to-side. A few alternative approaches are to use the specially made mylar type "solar eclipse glasses" (*not* sun-glasses worn at the beach) or a large piece of shade #14 or darker welder's glass. The glass is safe for naked eye sunspot watching since it removes IR and UV light, while the #14 shade darkens the solar intensity sufficiently. Welder's glass, depending on its make-up may tint the Sun green; a telescopic objective filter will as a rule give the Sun a hue of white, blue or orange.

For sunspots to be visible several criteria must be met. The size of the spot is important. For instance, in bright sunlight the pupil of the eye contracts to about 1.5 mm. Dawes Limit (116/aperture in mm = resolution in arc second) tells us that a spot for discernment with the typical unaided eye must subtend an angle of about 77 arc sec. Other factors affecting visibility include: seeing conditions, the position of the Sun, and the inherent resolution of the observer's eyes.

Poor seeing can deform and reduce the contrast of a spot making it increasingly difficult to detect. The altitude of the Sun plays a role because the light from a low placed Sun has more layers of turbulent air to transverse. The general rule of thumb is that the greater altitude at which you find the Sun, the better.

Individual acuity (the capacity to make out fine detail) plays another big role in identifying naked eye sunspots. Think about legendary observers like Edward E. Barnard or S. W. Burnham and the acuteness demonstrated in their astronomical discoveries. An individual's ability to distinguish subtle differences in contrast and grasp optimum resolution with their eyes may ultimately be the influential factors for seeing a naked eye sunspot.

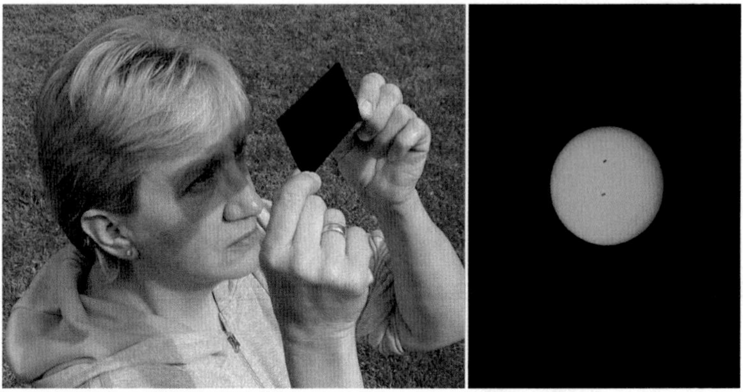

Fig. 2.2 Viewing naked eye sunspots

While of little scientific value, naked eye observations are a test of your eyesight and a sidebar to conventional solar observing. It is a fun activity that adds a "remember when moment" to a spectacular sunspot.

Figure 2.2 pictures a solar observer glimpsing several sunspots through a piece of #14 welder's glass. On the right side we see what the observer sees, a filtered view of the solar disc and in this case two tiny black umbrae.

Limb Darkening

Your first telescopic view of the Sun will present the globe as a unique three-dimensional orb. The central region of the Sun looks brighter than the outside edges. This is an interesting effect known as limb darkening.

Limb darkening is a visual perception caused by our ability to view slightly deeper into the center region of the Sun, where it is definitely hotter and therefore brighter. The regions nearer the limb appear progressively darker because they are seen at an increasing angle, where the cumulative amount of gas has acquired greater opacity, permitting only the upper photospheric layers to be discerned. These upper layers have a temperature near 5,000 K, the deeper view at the Sun's mid-region closer to 6,000 K.

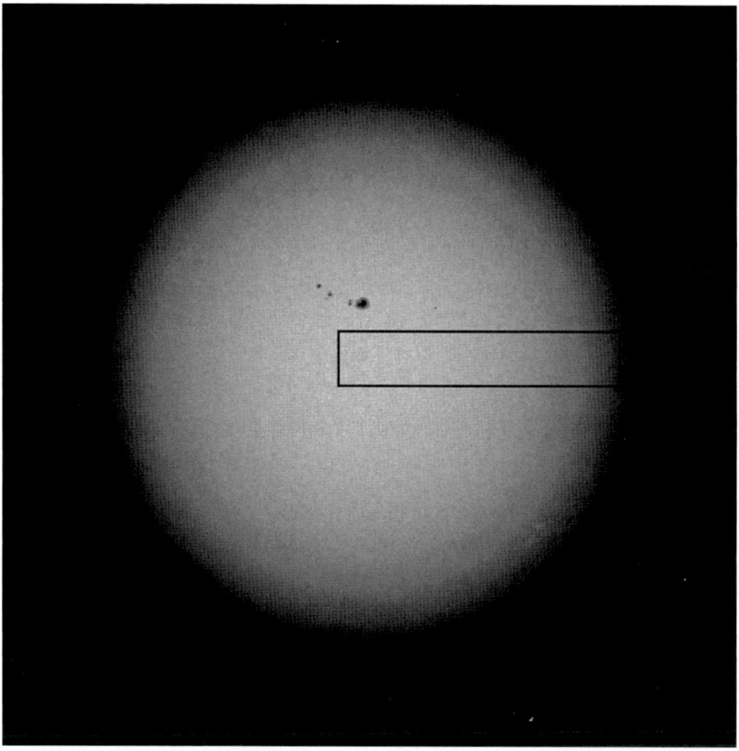

Fig. 2.3 Limb darkening on the white light Sun evident by the difference of density within the box. Notice how the center is brighter than the limb region

If our eyes were infrared sensitive, darkening would be essentially invisible. In ultraviolet light the darkening is still noticeable. However, limb brightening, a reversal of limb darkening, takes place at the far ultraviolet wavelengths.

Photometric studies by an observer at various points from the Sun's center to the limb, and at a variety of wavelengths, might provide some interesting data pointing out the differences in photospheric opacity. This experiment can be accomplished by positioning a photometer to measure filtered light near the western edge of the Sun and then allowing diurnal drift to scan the disc producing a brightness profile. A graph of the photometer's output will provide a clear picture of the limb darkening process (Fig. 2.3).

Solar Activity Cycle

If you study the Sun long enough (several decades) it becomes apparent that it experiences periods of increased and decreased activity. During several years the photosphere will be constantly churning with literally dozens of sunspots, while other times hardly anything is seen for weeks on end. Similarly during these periods the chromosphere can be ablaze or rather quiet. The time having the greater amount of activity is called the solar maximum, or solar-max for short. Likewise when little or no activity is occurring we call that the solar minimum. The stretch between a max to minimum and back to a max works out to be anywhere between 8 and 14 years, the average working out to around 11.1 years. Hence, this period is called the 11-year solar activity cycle.

The German amateur astronomer and pharmacist, Samuel Heinrich Schwabe, discovered the solar cycle in the early to mid-1800s. Schwabe had developed a clever plan for locating a suspected planetoid thought to orbit between the Sun and Mercury. In essence the strategy was to be on the lookout for the mysterious planet crossing the face of the solar disc. At the heart of the plan was the necessity to maintain records of all intervening sunspots, so they could be eliminated as questionable quarry. After some time he found no planet, but he did discover a periodicity to the appearance of sunspots. Schwabe thought the cycle to have a length of about 10 years, but additional observations refined the interval to the current values.

Every new cycle is given a serial number distinguishing it from each successive solar cycle. Solar astronomers established that cycle number one would begin in the year 1755 and end about 1766. Thereafter, each following cycle has been well tracked, providing useful information on the activity of the Sun throughout recent history.

New solar cycles begin with sunspots forming in high solar latitudes, near to the polar regions. As a cycle progresses, the amount and frequency of sunspots increase, then slowly as time passes newer spots will appear closer to the solar equator. When graphing the heliographic latitude of sunspots against time, a unique pattern known as a butterfly diagram becomes evident, see Fig. 2.4. To reach the activity crest an average of 4.8 years transpires.

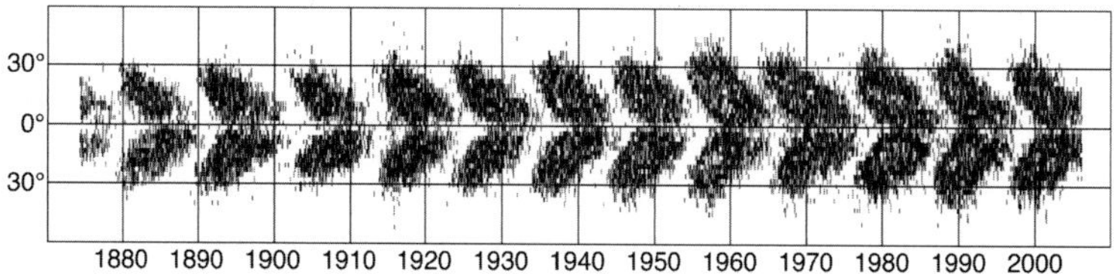

Fig. 2.4 Butterfly diagram (Courtesy of D. Hathaway/NASA)

During solar maximum the Sun acquires a daily sunspot count in excess of 200 on some days with several groups producing solar flares of various intensities. In the chromosphere prominences are frequently visible about the solar limb and across the disc. Solar minimum is gradually attained after an additional 6.2 years of slowly declining activity. The prediction of a cycle's strength and duration is a difficult task for solar astronomers. Many of the facets that play into a typical solar activity cycle are yet to be understood.

It's worth noting that bi-polar sunspots have positive and negative polarities whose magnetic alignment is determined by the hemisphere in which they appear. If the leading spot of a bi-polar group is of positive polarity then the following spot will be negative. This will be true for all sunspot groups of that hemisphere. Oddly, all bi-polar groups in the other hemisphere will be aligned with a reversal of that polarity, the leading spot becoming negative and the following positive.

Even more unusual is the phenomenon that takes place when a new solar cycle begins. A complete reversal of the magnetic field of the Sun happens, so that each hemisphere's sunspots have a polarity opposite that of the previous cycle. Some solar astronomers use this "field reversal" as the telling point to assess when a new solar cycle has begun.

In order for the magnetic polarities to return to their previous state, two solar cycles will have had to occur. The reversal and return of the Sun's magnetic field to its original state is called the magnetic cycle and averages 22 years.

Amateur observers with a preference for sunspot counting or positional recording of spots and groups can use their data to construct their own

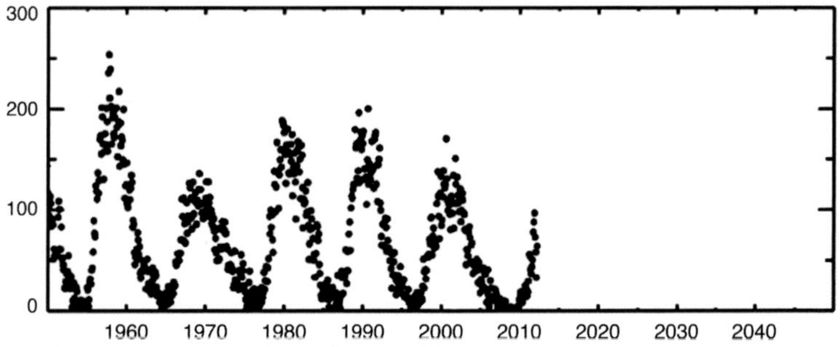

Fig. 2.5 Solar activity cycle plotted through the years (Courtesy of D. Hathaway/NASA)

butterfly diagram or activity cycle graph (see Fig. 2.5), detailing the yearly progress of Sun activity. These visual presentations correspond to the real solar waning and waxing experienced at the telescope.

Granulation

Granulation is a low contrast, high resolution feature, which covers the entire photosphere of the Sun. Described variably as rice grains, rice pudding, kernels of corn, multi-sided polygons, or other geometric patterns, perhaps the most intriguing description is that of an "orange peel". Take a large, thick skinned orange, hold it at arms length and slowly rotate it on your fingertips, and you will have an exaggerated representation of how the feature appears on the Sun. A low-resolution view almost always will show granulation as having a rounded appearance.

Granulation is composed of very small (approximately 1–5 arc sec) elements, each called a granule. The surface of the Sun finds itself covered with approximately 2–3 million granules at any given moment. This feature is less visible near the limbs; attribute that to foreshortening because of its depth within the photosphere. However, granulation does become markedly easier to spot near the middle of the Sun's disc. There is a definite lack of uniformity to the brightness of granules. Some appear dull and others by comparison somewhat brighter. Figure 2.6 by solar photographer Art Whipple illustrates this characteristic well.

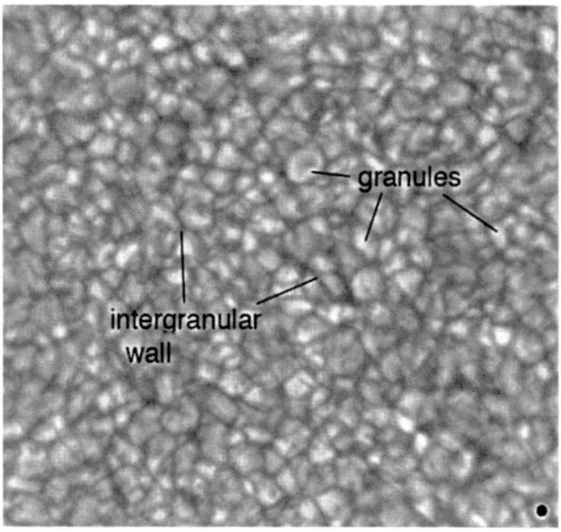

Fig. 2.6 Granulation near the center of the Sun's disc. Individual granules and the intervening intergranular walls are *marked*. The *small black dot* in the *lower right corner* is 1 arc sec in diameter (Image courtesy of Art Whipple)

A granule is the top of the column of gas (plasma), rising from deep within the convection zone. The plasma ascends to the solar surface subsequently to release its energy. Brighter granules have more stored energy than the duller counterparts. After releasing the energy, the plasma cools, becomes darker and flows back into the solar interior along what is called the intergranular wall. This wall or lane forms a boundary separating granules; the wall defines each granule's unique shape. The lifetime of the typical granule is 5–10 min, greater with the larger specimens. A change in shape is visible after only a minute or two of observation; large granules often form out of the fragments left behind by the breakup of previous large granules. Smaller granules tend to fade away. All in all, the majority of granules perish by either splitting apart or disintegrating. A few will disappear by merging with neighboring granules.

A noteworthy observation is that granules located near the penumbra of a sunspot will have a compressed diameter as compared to their outlying neighbors, the result of a sunspot's magnetic influence.

Granulation is difficult at best to observe. The short life span combined with a necessity of excellent seeing conditions make studies for the typical solar observer a challenge. During a time of superb seeing, visual observations will be enhanced by following these recommendations: look near the center of the solar disc, use a supplementary green (i.e. Wratten #58) filter or broadband interference filter transmitting near 540 nm, consider that higher magnifications (75× or more) may be required to resolve individual granules, and lastly apply extreme patience and expect only fleeting glimpses unless the seeing is abnormally good. The successful granule observer will possess sensitivity to a mild shift in shape, size, and brightness during the short lifetime of the feature.

Photographic methods are by far the preferred means of serious granulation studies. Video recording captures the infrequent instants of fine seeing, and then later the photographer can pluck the sharpest images from the video. These superior samples are then assembled into a series of images depicting the feature's evolution or used for the creation of a time-lapse movie showing the changing morphology.

Supergranulation is a large-scale pattern found in the photosphere of organized cellular structure. Each structure contains hundreds of individual granules with a supercell diameter of roughly 30,000 km and a lifetime from several hours to about a day. The observable side of the Sun shows something like 2,500 of these supergranulation cells. The feature is virtually invisible to the white light observer; however it can be detected in Ca-K and H-alpha light. The chromospheric network overlies this feature.

Faculae

Decidedly easier to observe than granulation, even in less than ideal conditions, are the wispy, cloud-like faculae (singular is *facula*). Facula means a bright point or small torch; the vast majorities of examples however are not at all point-like, but rather are venous streaks, having a brightness slightly greater than the surrounding photosphere. Faculae tend to surround sunspots, developing a length parallel to solar rotation and having a width of several heliographic degrees or more.

To understand the cause of faculae, it is helpful to comprehend that sunspots are the result of strong magnetic flux stifling convection from below the photosphere. Facular regions, too, are the result of magnetic fields active in the photosphere, but while sunspots arise where a strong field is present, a magnetically weaker field produces faculae.

One characteristic of stifled convection is an apparent depression given the encompassing solar surface. An obvious example of this depression is visible when a symmetrically shaped sunspot approaches the solar limb. The spot gradually assumes a cavity-like appearance, the so-called Wilson effect. In reality the Wilson effect is the result of magnetically influenced gas becoming more tenuous or transparent, thus permitting an observer an opportunity to see a bit deeper into the photosphere. This is what produces the illusion of a saucer-shaped depression. The weaker magnetic fields associated with faculae likewise produce a fissure effect resulting in light emerging through the "sidewalls" of the depression. Scattered light from this sidewall leakage causes the faculae to be brighter and nearly 100 K warmer than the surrounding photosphere.

All sunspot groups are linked to faculae, but not all faculae have attending sunspots. Faculae are a precursor to developing sunspots and can experience a life lasting up to several rotations of the Sun. It is therefore quite possible to observe "orphaned" facular regions, particularly if the facula has developed outside the sunspot zone of ±35° solar latitude. The zone for faculae development is wider than the sunspot zone, extending to about ±50° heliographic latitude.

When faculae occur outside the sunspot zone and expressly near the polar regions of the Sun (±60° to 90°), they are known as polar faculae. Polar faculae differ from ordinary faculae in area and lifetime. Small granular sized points or elongated flakes, polar faculae last from only minutes to a day or two at most. During the minimum of an 11-year solar cycle, polar faculae are seen more frequently. Observing the smaller area of polar faculae necessitates a telescopic aperture of 100–125 mm, used during moments of fine seeing conditions.

Solar projection is not the best method of observing this feature. The low contrast of a facula makes it difficult to spot on a projection screen. Direct observation is the preferred method of viewing both faculae and polar faculae. A mylar type objective filter transmitting in the blue region of the spectrum is ideal. Not having that accessory, try using the Baader AstroSolar™ Safety Film coupled with a light green (Wratten #56) or light blue (Wratten #80B) eyepiece filter. The supplementary filter boosts contrast of the facula, making it easier to distinguish. Photography is enhanced with even deeper green filters such as the Wratten #58 or a Wratten #61, and for locating faculae furthest from the Sun's limb try a Wratten #47 (deep blue) (Fig. 2.7).

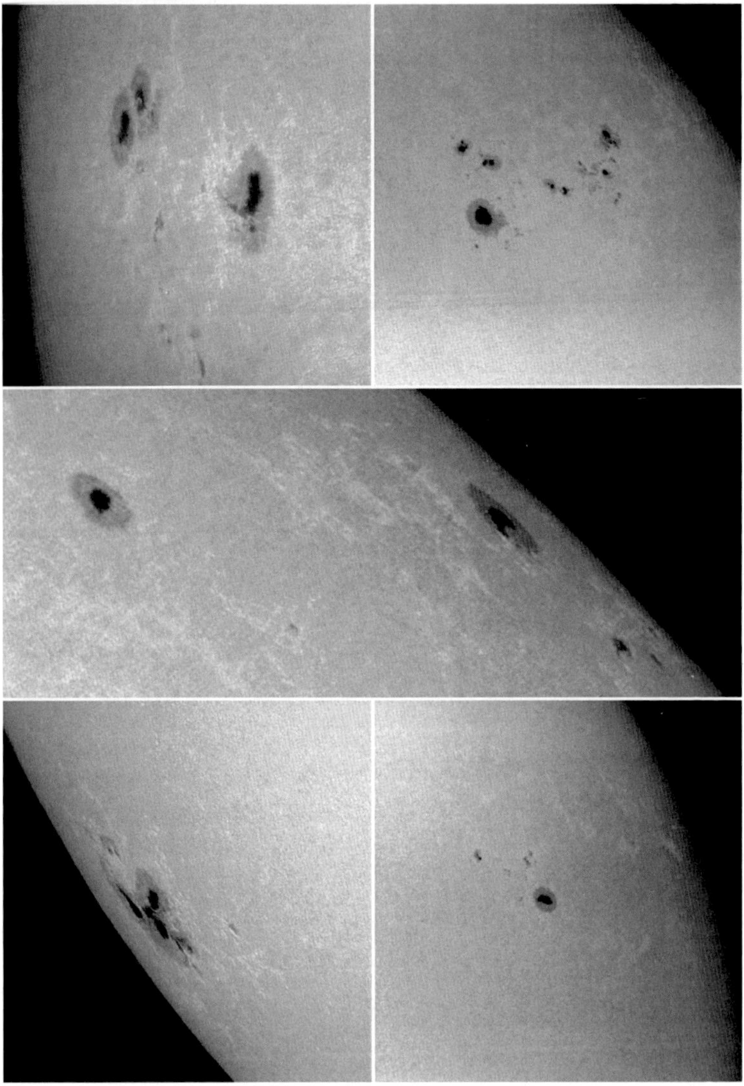

Fig. 2.7 Examples of faculae appearing near the solar limb and associated with sunspots. Faculae are wispy, cloudlike features, slightly brighter than the photosphere

Bright Points

Appearing between granules at the intergranular wall and sometimes on the very edges of granules are the so-called bright points. These miniscule features (average size 0.25 arc sec) result from intensive but small flux tubes (300 km) poking through the photosphere that allows an inspection of the solar interior. The name is quite descriptive of the visual appearance of the feature, bright point-like dots randomly located on the darker lanes between granules.

A group of bright points, forming a string or chain upon the intergranular wall, are known as *filigree*, observable in the solar continuum but more frequently at the G-band (430.5 nm) or in H-alpha. In the literature, bright points are often referred to as G-band bright points because of the increased contrast given them at that wavelength.

Pores and Voids

At times the intergranular wall may appear duskier than normal, or a granule may look darker than surrounding granules, or even missing; this is customary because of a natural variation in brightness with granules. Identifying brightness differences between granulation and other solar features called pores and voids can be a challenge. There is also an art to distinguishing a pore from a sunspot, because any sunspot lacking a penumbra may possibly be a pore.

Consider the following characteristics when separating pores from sunspots. Pores are defined as tiny features with a diameter usually between 1 and 5 arc sec, averaging about 2.5 arc sec. For the most part pores are symmetrical in shape. They are darker and bolder than granular material though not as dark as the umbra of a well-developed sunspot. The intensity of most pores is 0.2–0.4 of the surrounding photosphere. The intensity of the photosphere is given a numerical value of 1.0 and is commonly allocated the term, I_{phot}.

Size has a direct relationship to longevity, and longevity a relationship to a pore developing into a sunspot. Large pores may last for several hours, while smaller examples can form and then dissolve in a few minutes. Pores with a size greater than 5 arc sec have the best chance to grow into a sunspot.

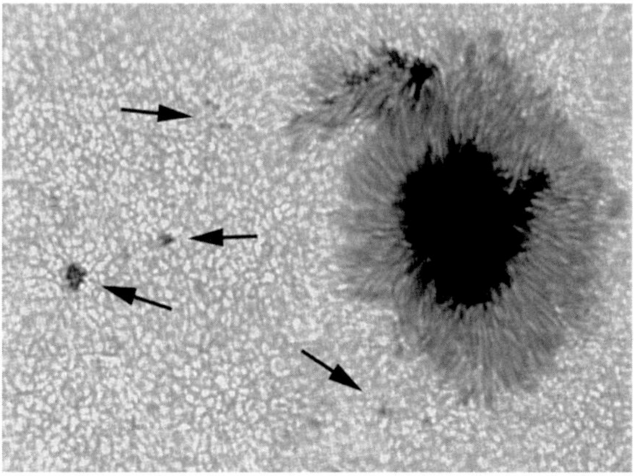

Fig. 2.8 Two features sometimes difficult to tell apart, dark granules and pores, accompany the larger sunspot on the *right*. Dark-appearing granular material is *arrowed right* while several pores are *arrowed left* (Courtesy of Eric Roel)

While many times found near existing sunspots, pores can also be located in isolated faculae. Like sunspots, the formation of pores is the result of magnetic flux tubes extending upward through the solar interior and restricting convection.

The void is not a pore; it is a circumstance where granulation is actually found to be missing. Again these are dark areas about 1–5 arc sec, with intensity near 0.7 I_{phot}. While the typical pore is round, a void may have an irregular shape, which can in a few minutes change brightness as it begins to fill with a granule (Fig. 2.8).

Sunspots

Revealed several thousand years ago by Chinese astronomers, a sunspot is still the most observed feature of the white light Sun. The first impression of many new observers is that a hole or crater has blemished the near perfect appearing photosphere. Repeated observations indicate that these blemishes go through a development process that includes several fascinating phases.

Sunspots are birthed through a generally repeatable pattern in white light that follows this basic scheme. Bright faculae will develop about 7–14 days before a sunspot appears. A number of pores will begin to cultivate within the facular region. The majority of these will be granular sized, ultimately to decay and disappear, however some pores will become larger and darker, becoming as dark as an average sunspot's umbra. This is called an umbra spot, a point where development is sometimes arrested and decay occurs in a short time. The umbra spot can continue to experience growth and a rough appearing penumbra likely will appear. This penumbra could become intricate, harboring islands of dark umbrae containing slightly brighter umbral dots within their interior. Now we have a fully developed sunspot, most likely accompanied by similar shaped features, each experiencing its own evolutionary cycle.

Sunspot structure is often the focus of the amateur solar observer. Whether the interest is in the changing appearance of a feature (morphology), or in maintaining an index of solar activity (sunspot counting), observing sunspots and their associated phenomena provide an excellent opportunity for increasing your understanding of how the Sun works (Fig. 2.9).

Umbrae

Umbra (the plural form is umbrae) is the darker central region of a sunspot. To the eye an umbra appears dark only because it is in contrast against the surrounding, brighter photosphere. In reality the brightness of a typical umbra is in excess of any star in the night sky.

Detailed examination of an umbra reveals that the interior is not the flat charcoal colored smudge it presents. Rather an umbra is made up of shadowy granules, tiny bright points, and material with intensity somewhere between the dark granules and bright points called umbral dots.

Average intensity in the umbra interior is on the order of about 0.1 of I_{phot}. Variables affecting the brightness of any umbra include the seeing conditions, light scattering and the resulting loss of contrast, as well as the chosen wavelength of observation. A direct relationship exists between brightness and temperature of an umbra, therefore the temperature of an

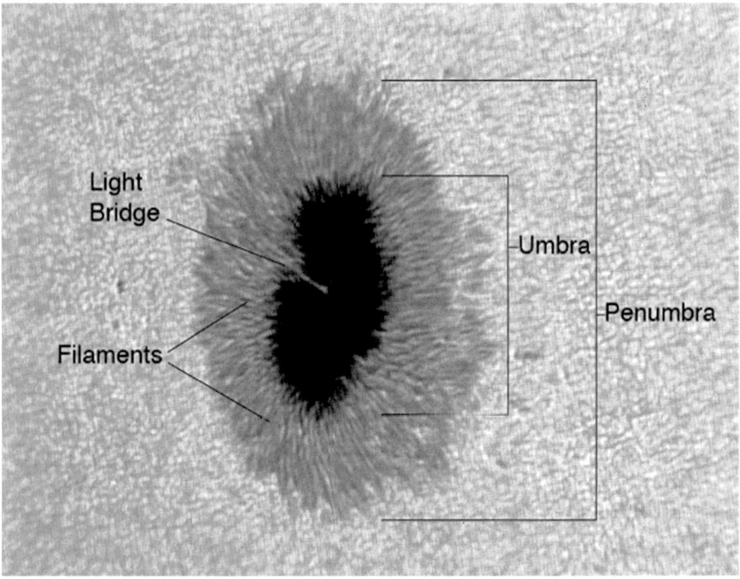

Fig. 2.9 The general anatomy of a fully developed sunspot includes, as seen in this photo by Eric Roel, *a dark interior* called the umbra and a surrounding donut of *lighter gray*, the penumbra. *Thread-like filaments* compose the penumbra and a small light bridge is just beginning its trek to divide the umbra into two segments

umbra can be calculated from the Stefan-Boltzmann law: $I/I_{\text{phot}} = (T_e/(T_e)_{\text{phot}})^4$. Based on a given photospheric temperature, $(T_e)_{\text{phot}}$, of 5,780 K and an I_{phot} of 1.0, the typical umbra computes a temperature near 3,300 K.

There can also be a variation in the umbral color. At first glance the umbra may give an impression of black or a very dark gray. Examine the umbra closer with a filtration system neutral in its spectral transmission (i.e. Herschel wedge or Baader Astrosolar film) and subtle shades of color may become evident. Black and shades of gray may graduate into a dusky deep reddish-brown hue. The stronger the magnetic field of the sunspot, the darker an umbra appears. Past observations have revealed that umbrae appear darker during sunspot maximum than at the sunspot minimum (Fig. 2.10).

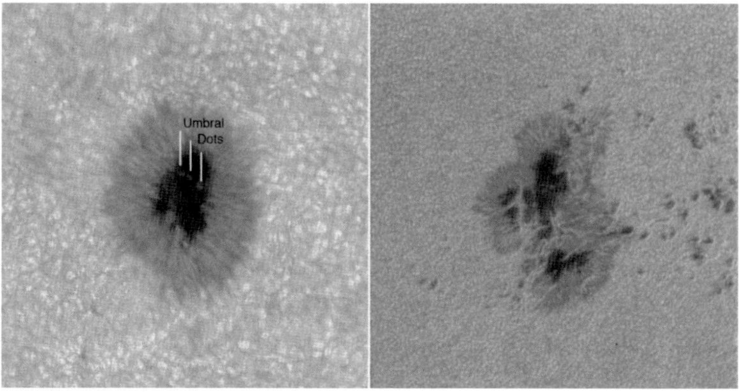

Fig. 2.10 The *left-hand image* obtained with a 14-in. aperture f/4.6 Newtonian telescope illustrates fine interior details of an umbra. Several streaks are visible and at the foot of the *three markers* are umbral dots. Art Whipple is the imager. *To the right* is another high-resolution photo depicting the appearance of numerous umbral dots visible near the photo's center (Courtesy of Eric Roel)

Umbral dots (UD) are round features with a diameter about 0.2–0.5 arc sec and intensities on the order of 0.1 I_{phot}. Slightly brighter and therefore hotter than the surrounding umbra, UDs resemble clumps of granular material located within the umbra. The classification of an umbral dot is determined by its umbral location, central or peripheral. A peripheral UD is located near the surrounding penumbra; a central UD is positioned nearer the middle of the hosting umbra. Peripheral dots have a habit of gradual movement toward the central region of the umbra where they are apt to remain stationary features. Umbral dots are conjectured to be the result of minimally active convection within the umbra. UDs are difficult to observe and like most low contrast solar forms necessitate excellent seeing conditions and an instrument capable of resolving their diminutive size.

The amateur astronomer desiring to study such sunspot phenomena can apply a couple different photographic techniques. The first utilizes deep exposures of the umbra to reveal detail frequently missed by the visual observer. Another technique, the creation of an isophote map of an active region illustrates sectors of equal density that translate into various

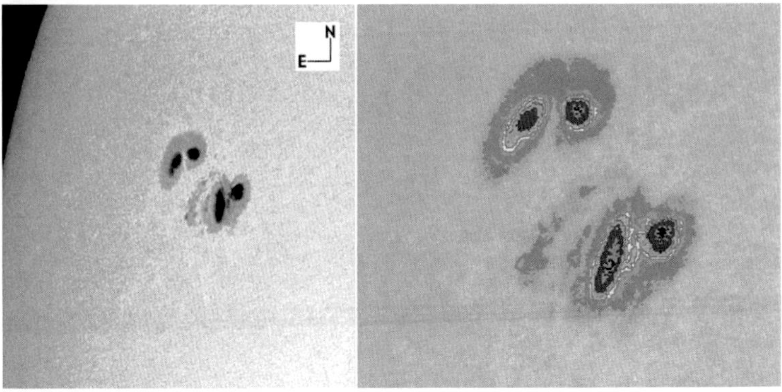

Fig. 2.11 Isophote map created using NIH-image software. In this map the conventional photo *on the left* had density slices extracted and recombined to form the image *on the right* showing areas of similar density. The leading umbras of each bi-polar group contain the core or region of greatest density and therefore the coolest temperature

regions of differing temperature. An isophote map also indicates the point of coolest temperature (greatest magnetic strength) in the umbra, a zone identified as the sunspot's core.

Briefly let's touch on how these methods are accomplished. A deep exposure is attempted by first masking out the photosphere and the surrounding penumbra and permitting only the light from the umbra to reach the detector. Give several times the standard exposure to reveal the granular nature of an umbral interior. Excellent seeing conditions are of course necessary for the success of this technique. An isophote map is easily created with readily available software applied to medium through high definition digital images. Recommended software includes the *NIH-Image* (Macintosh version) or *ImageJ* (Windows PC version), both available as freeware.

The basic procedure is to create numerous individual images representing "slices" of photographic density from the original photograph, colorize each slice, and then recombine the slices, forming a descriptive layered map. Features easily picked out from a map include: regions of similar temperature, the sunspot core, inner and outer bright rings, as well as weaker light bridges. Details on creating an isophote map of a sunspot can be found in the Springer book, *The Sun and How to Observe It* (Fig. 2.11).

Penumbrae

Sunspot penumbra has been likened to a halo, at lower resolution seen as a lighter gray outer region surrounding the darker umbra. Rudimentary penumbra often forms from the intergranular material adjoining newly developed umbra. Any spot possessing a large umbra will have developed a penumbra which evolves structures of dark penumbral filaments that radiate about the umbra like fine threads, 0.2–0.3 arc sec in width. These threads are magnetic in nature and have a similarity to granules in their convective characteristics. Between the dark threads are brighter comet-shaped regions called penumbral grains. Superb seeing conditions and an instrument capable of resolving better than 1 arc sec are required to distinguish penumbral filaments and grains.

Penumbral grains are found to move outward, away from the umbra with speeds of 0.5–2.0 km/s^{-1}. This movement is called the Evershed effect, discovered in 1909 by John Evershed, as he observed a Doppler shift in the spectral lines of sunspots. Penumbral filaments possess a horizontal alignment following the sunspot's magnetic field lines, which emerge vertically from the umbra and then become inclined at an angle of 25°–30° to the photosphere near the outside edge of the penumbra.

A mature sunspot will usually have a penumbra that is symmetrical in shape. Less often seen is an irregular penumbra, which has been mutated by complex magnetic fields. An irregular penumbra will inundate the sunspot group with filaments of varying widths throughout. Independent islands of penumbrae may appear separated from the umbra. This condition is somewhat infrequent and rarely lasts longer than a day.

Within highly developed penumbra may be found islands of dark umbral material that are only a bit larger than pores. Regions of material as bright as or brighter than the photosphere may also be found. These dark and bright regions are known to undergo rapid changes, and ought to be observed closely. Occasionally, a bright region may become elongated and fade, transforming itself into filaments or growing larger into what is known as a light bridge (Fig. 2.12).

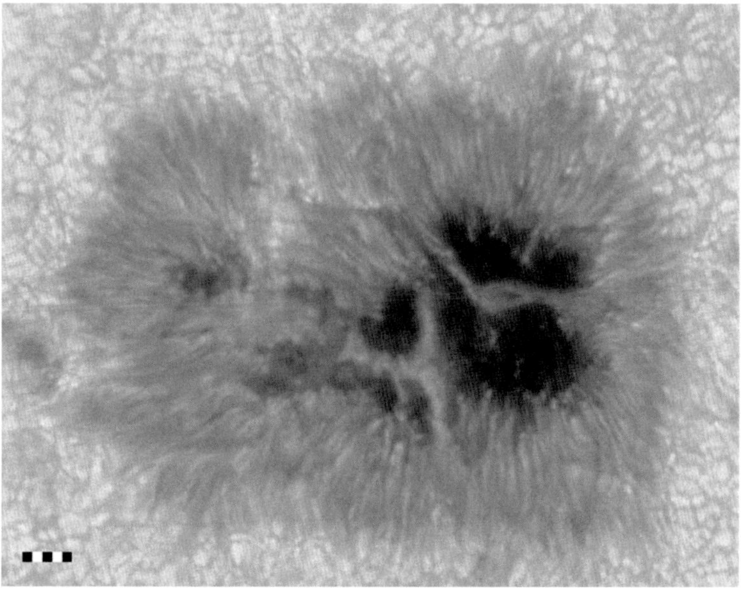

Fig. 2.12 High Definition solar photography. Notice the dark central lane of the oval structure in the light bridge that divides the larger umbral region. Umbral dots and small-scale granulation in the more developed light bridges are visible. In the penumbra the *dark radial threads* are filaments and the *white "comet-shaped"* features are penumbral grains. The *squares* in the *lower left corner* represent a scale of 1 arc sec (Photograph courtesy of Art Whipple)

Light Bridges

A light bridge is loosely defined as any material brighter than an umbra that also divides the umbra, sometimes even dividing the penumbra. The perception is frequently given that the photosphere has started to flow into the sunspot, somewhat like a river coursing a new path. Three varieties of light bridges may be categorized: the penumbral island, the streamer, and the familiar classical light bridge. Over time a light bridge can transition from one category to another, making it difficult to classify. A sunspot group may contain an assortment of light bridges at any one time.

Older, mature sunspots often contain a well-developed, thickly proportioned light bridge (LB), the typical classical variety. In some venues this bridge is referred to as a photospheric light bridge because of its structural similarities to nearby photospheric granules. It is important to note that granules surrounding sunspots are smaller than those found in the quiet photosphere. Likewise the photospheric LB is made up of similarly sized compacted granules. The photospheric LB is often brighter than the photosphere, with a lifetime of 1–14 or more days.

Younger sunspots many times develop a light bridge that is thin, streaky, and intense. They are known to mature into a semi-complex network that resembles a "lightning strike." This is the so-called streamer. Angularly a streamer is a narrow 1–5 arc sec in width. Like the classical LB, a streamer also has a granular nature, though the grains are somewhat narrower than photospheric granules. Some granules may appear elongated and bright, reminding the observer of a penumbral filament. At other times a streamer appears to be composed of dots, similar to but brighter than an umbral dot. Streamers have an erratic lifetime of several hours to several days.

Penumbral islands are located in the penumbra of a sunspot. This is an independent area of material (also granular in nature) that is brighter than the surrounding penumbra. Appearance of the feature can deviate from an irregular patch to a streak of any size and shape. Penumbral islands are known to last for several hours to several days.

Granular structure is a common thread between all classes of light bridges. Figure 2.12 captures this characteristic well. In the late nineteenth century the astronomer Janssen was the first to note the granular structure of light bridges (Fig. 2.13).

There is no strict "rule-of-thumb" for light bridge development because a LB may well begin to grow, and then at any stage of growth become quiet and possibly reverse its course. However, there is a guideline to the development of the classical photospheric light bridge that follows this basic outline. In a sunspot lacking a light bridge, bright penumbral filaments will begin to enter the umbra, many times from opposing sides of the umbra. Individual granules will begin to separate from the filaments crossing the inner umbra, forming a streamer and completing the connection of the opposing sides of penumbra. As the LB connection grows, remaining bits

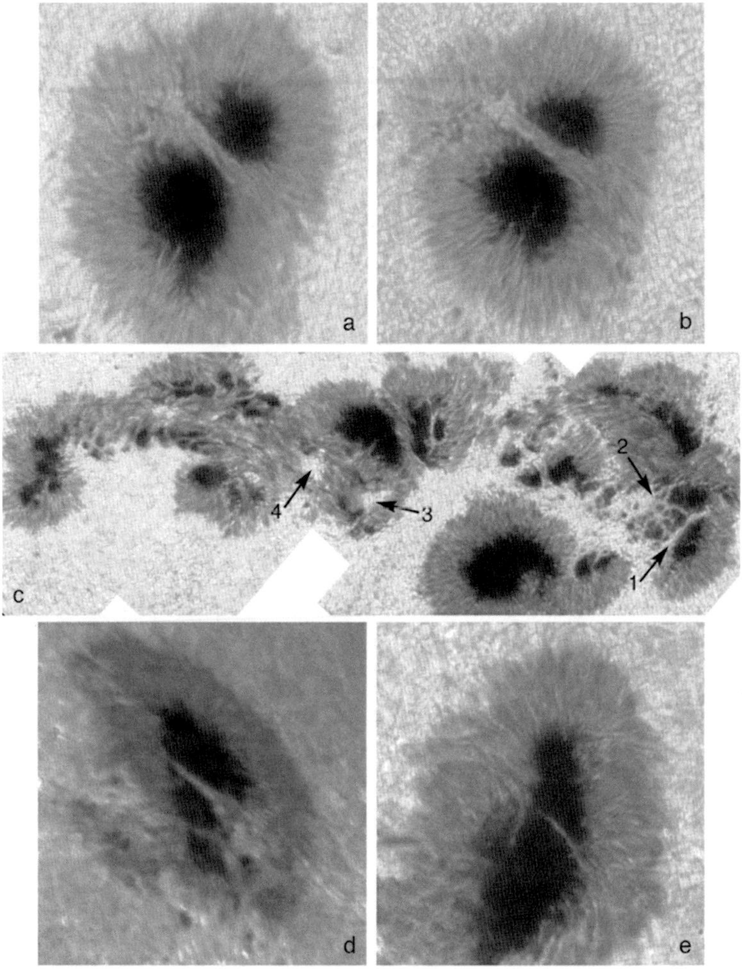

Fig. 2.13 Classic light bridge development over a 24-h period is visible in photos **a/b** of active region 11,289 from the 10th and 11th September 2011. Notice how the penumbra filaments appear to have grown a "finger" across the umbra. The granular nature of the light bridge is obvious in the original photo within the *left area* of the LB. The *middle photo*, AR9393 from 28 March 2001, contains examples of all three types of light bridges. Beginning from the *right*, **c1** is a classic photospheric LB, streamers can be found at several locations, but prominently are visible at **c2**. Penumbral islands are at **c3** and **c4**, near the center of the photo. The example on the *lower left*, AR11109 from 2 October 2010, has a streamer midway in the umbra developing into a classic light bridge. The umbra of AR10110 supports a dividing thin streamer on 21 September 2002 (All photos courtesy of Art Whipple)

of filament or umbral material continue to decay leaving behind a classic light bridge dividing the sunspot umbra.

Studying the development of a light bridge can be a relaxing activity. Several days can pass before the bridge has completely divided an umbra. A light bridge at times is born with a new sunspot, and can stay throughout its lifetime.

All light bridges display a wide range in intensity or brightness. Some can be so weak that they are practically invisible, only showing up best in digitally enhanced photographs, while other examples are somewhat brighter than the surrounding photosphere. Viewed near the solar limb a light bridge is time and again particularly intense, outshining any nearby faculae.

Use of a Wratten #21, #23, or #25 red filter will darken the umbral area of a sunspot causing an increase in the contrast of light bridges. Any of these filters are also helpful for discerning details in the structure of penumbrae, such as knots and streaks (Fig. 2.14).

Bright Rings

The inner and outer bright rings of a sunspot are a misnomer; the rings are not bright in the sense that they greatly outshine other features. Rather the rings when detectable, and all sunspots have rings, are only ever so slightly brighter than their immediate surroundings.

The inner bright ring is situated between the outer umbra and the inner penumbra, a thin region occupied by the comet-like heads of the penumbral grains (see Fig. 2.12). The inner bright ring is surely an increase in intensity found in the penumbra, extremely difficult to perceive, but traceable photometrically.

Beyond the penumbra of a sunspot in the immediate vicinity of the photosphere is found the outer bright ring. Having an intensity of approximately 1.0–1.1 I_{phot} the outer bright ring generally is less thick than the sunspot's penumbra. The outer ring is brightest near the penumbra and decreases

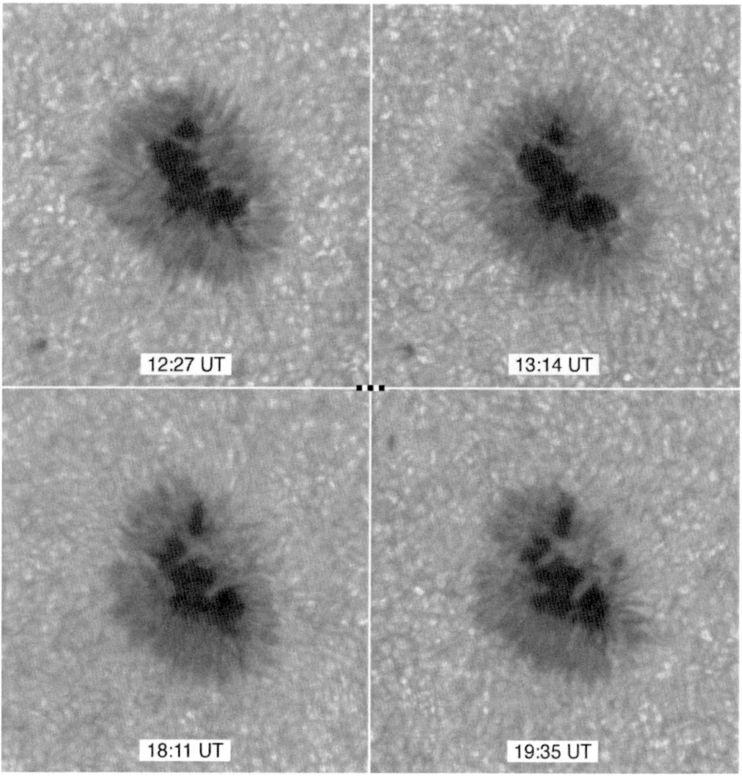

Fig. 2.14 Light bridge development over a 7-h period on 31 August 2011 is visible in these four photos. Fingers are growing from both sides of AR11279's penumbra, soon to connect after the 19:35 UT photo. Each *square* at the *center* represents 1 arc sec (All photos courtesy of Art Whipple)

with distance from the sunspot. This feature is best visible in blue or violet light, and rarely is noticeable in orange or red. Productive studies have been also performed in monochromatic Ca-K light. Rings *may* be more difficult to observe near the limb owing to the presence of the yet brighter faculae. Bright rings typically assume the shape of the host sunspot and can be only partial ill-defined rings, or entirely closed.

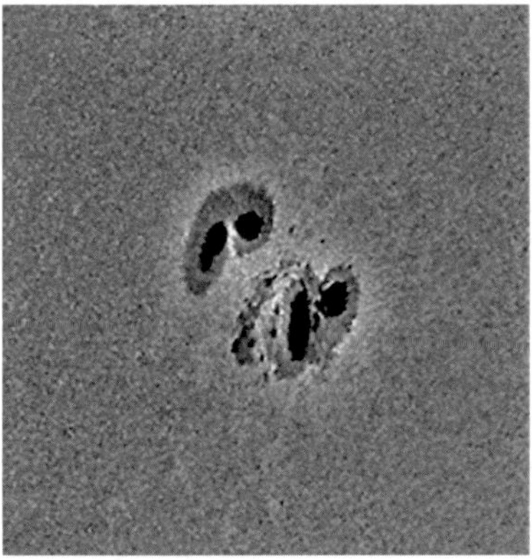

Fig. 2.15 The image from Fig. 2.11 has been digitally enhanced to show the outer bright rings surrounding each bi-polar sunspot

Bright rings seem to have a strong relationship to energy transport within the sunspot. Below the surface, convection is stifled by the sunspot's magnetic field, it is conjectured that part of this energy is distributed to the areas occupied by the rings, resulting in their increased brightness and temperature (approximately 75 K warmer than the quiet photosphere) (Fig. 2.15).

Wilson Effect

Alexander Wilson in the year 1769 discovered an interesting property of symmetrically shaped sunspots. When one of these spots approaches the solar limb the width of the penumbra, relative to the umbra, on the side facing the center of the Sun seems to become narrower than on the side facing the limb. This property, known as the Wilson effect, is perceptible with a round sunspot, but not so much with an asymmetrically shaped spot, which may show a completely opposite effect.

Observing the Wilson effect is best accomplished as the sunspot is nearing the west limb, although a reverse occurrence can be observed on the eastern limb. Several days before rotating out of sight, a spot will begin to appear narrower due to foreshortening. As the penumbra and umbra both become thinner, the umbra will seem to move closer and closer to the penumbral side nearest the Sun's center. A point will be reached where the trailing edge of the umbra indeed will touch the edge of the trailing penumbra; this is denoted as the "contact state". Continued solar rotation causes the umbra to narrow further and eventually disappear.

Wilson's conclusion to the cause of this effect was that sunspots are cavities in the Sun's surface with differing depths. The modern theories are not so far from Wilson's thoughts. The Sun is a gaseous body with the outer layers being of less and less density. Therefore various parts of a sunspot have differing transparencies, meaning that their light originates at differing heights within the photosphere. The quiet photosphere has the least amount of transparency, a sunspot penumbra a bit more transparency, and an umbra the greatest amount of transparency. The greater the transparency, the deeper into the Sun the light originates. A diagram drawn showing equal optical depth of the photosphere, penumbra, and umbra gives an appearance likened to Wilson's cavity theory.

The contact state can be difficult to observe visually owning to normally less than ideal seeing conditions. Of course the larger the sunspot, the easier the Wilson effect is to see. A Wratten #21 or #25 filter is useful to darken the umbra and increase the contrast between umbra and penumbra (Fig. 2.16).

Sunspot Group Classification

Observe the Sun even momentarily, and you'll discover that a sunspot is not necessarily a solitary feature. A sunspot tends to develop a network of faculae, pores, and numerous other spots. These features, all magnetically associated, form a cluster called a sunspot group.

Groups typically, but *not always* are located outside 10° of heliographic latitude or longitude from other groups. This is known as the "10° rule" and is used by some amateur astronomers classifying groups in order to distinguish one from another. Occasionally, two clearly established groups may form within 10° of each another, and the rule has to be bent to

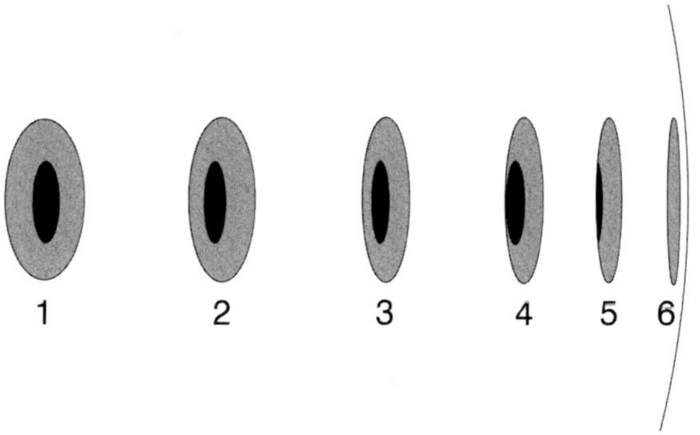

Fig. 2.16 The Wilson effect near the Sun's west limb. *Position 1* represents a spot some distance from the limb. As the spot gradually approaches the limb, the penumbra and umbra become narrower. The side of the penumbra facing the center of the disc will have less of a width than the side facing the limb. In *position 4* the umbra and penumbra eventually meet, this event is called the *contact state*. As the Sun continues to rotate, the umbra finally disappears from sight as the penumbra narrows further

accommodate the situation. Most times it will be obvious if several groups are clustered together, though on occasion making a distinction can be an art only mastered through experience.

Earlier in this chapter it was noted that sunspots go through classically repeatable life cycle patterns. A group develops through an observed blueprint that follows this course. In white light several pores may form a tight cluster confined within 10° of solar heliographic area. The pores will darken, becoming umbral spots (don't confuse them with umbral dots) sometimes gathered into two individual concentrations; this is classed a **bi-polar** sunspot group. After a short time, each concentration will develop a small sunspot. The two spots are identified as leading and following, the leading spot being the more westerly. Spot growth often terminates at this point with the dissolving of the newborn spots inside a few days. Should a group be particularly stable and continue to evolve, a penumbra will develop about the leader spot, and shortly thereafter around the other spots. The two sunspots will now begin to separate from each other by at least 3° in solar longitude, while rotating relative to the Sun's equator in an east–west direction. The leading (westerly) spot will have a magnetic polarity opposite that of the following (easterly) spot. A region or line separating the two opposing polarities exists, running in a roughly north–south direction that is known as the neutral line or magnetic inversion line. Sometimes sunspots rotate at a speed exceeding 1° an hour, wrenching the neutral line to a more east–west direction. This formation of the spots in sunspot jargon is known as **tongues**, and in appearance resembles the traditional yin-yang pictogram. This movement results in twisted magnetic field lines that precipitate events such as solar flares. Should the clumping of umbral spots consist of only a single concentration confined within a 3° area, it is called a **unipolar** sunspot group. On occasion an apparent bipolar group can contain numerous sunspots of differing polarities, all involved with a mutual penumbra. This assemblage is called a **multipolar** group.

Near the middle of the second week of growth a group typically maxes out in area and number of individual spots. Around this time, and for up to a month later the signs of decay begin to appear. Pores and tiny sunspots start to dissolve, while the following spot divides and fades until it disappears. As this is happening, the leading spot becomes symmetrical (rounder) in shape. Slowly the leading spot shrinks away, leaving any attending faculae that eventually disappear.

The transformations in this pattern can be identified by several classification schemes that solar astronomers have devised over the years. Max Waldmeir devised one well-known, and often used system in the early twentieth century. His Zurich sunspot classification system established nine classes labeled A, B, C, D, E, F, G, H, and J, each identifying the various stages of sunspot development. Excellent for cataloging a spot's lifecycle, the Zurich system fell short of the practical requirement needed in the mid to late twentieth century: the accurate prediction of solar flares. It is well known a solar flare can disrupt spacecraft and modern communication equipment, a concern of all people during the late twentieth century (Table 2.1 and Fig. 2.17).

Predicting when and where eruptive solar activity was most likely to happen took a significant step forward in the 1960–1970s when Patrick McIntosh created the extended version of the Zurich sunspot classification system. McIntosh altered the nine-class Zurich to a seven-division system. Two additional sub-classes were instituted describing the penumbra of a

Table 2.1 McIntosh sunspot classification system

Zurich class (modified)	
A	Individual spot, concentrated unipolar group, no penumbra
B	Bipolar group, no penumbra
C	Bipolar group, one spot with penumbra
D	Bipolar group, penumbra about both spots, length less than 10°
E	Bipolar group, penumbra about both spots, length 10°–15°
F	Bipolar group, penumbra about both spots, length greater than 15°
H	Individual spot, unipolar group, with penumbra
Largest spot within group	
x	No penumbra
r	Rudimentary penumbra surrounds only part of spot
s	Symmetric penumbra, 2.5° or less in north–south diameter
a	Asymmetric penumbra, 2.5° or less in north–south diameter
h	Symmetric penumbra, greater than 2.5° in north–south diameter
k	Asymmetric penumbra, greater than 2.5° in north–south diameter
Distribution within the group	
x	Unipolar group, class A or H
o	Very few or none, tiny spots between leader and follower
i	Many spots between leader and follower, none with developed penumbra
c	Many spots between leader and follower, one or more with developed penumbra

group's largest spot and the distribution of spots within the group. The added data derived from the three-letter McIntosh Classification System made it possible to more accurately forecast solar flares, giving amateur and professional astronomers an indication of when and where flares were likely to take place.

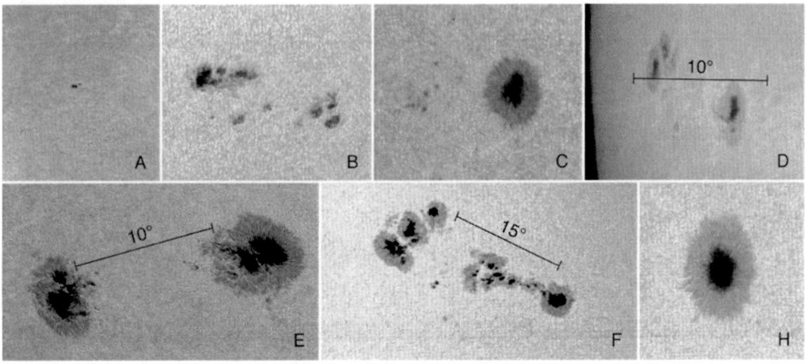

Fig. 2.17 Examples of the Zurich modified (McIntosh) group classes. Across the top is A, an individual spot with no penumbra, B is a bipolar group with no penumbra, C is bipolar, and in this case, the leading component has a penumbra. Groups D, E, and F have fully developed penumbra around the leading and following spots, however their separation varies as per the class (A, D and F from Jamey Jenkins. B, C, and H courtesy of Art Whipple. E is courtesy of Eric Roel)

Sunspot classification is a visual procedure requiring the astronomer to carefully inspect each sunspot group and conclude where that group falls within the descriptive language of the code letters. When all three areas of class are found, then the McIntosh classification for a sunspot group has been determined.

For example, upon inspection an observer concludes that a sunspot group is a bipolar group, has penumbra surrounding both sunspots, and has a length less than 10° of solar longitude (D), that the penumbra around the largest spot in the group is asymmetric with a north–south diameter less than 2.5° (a), and lastly, there are many smaller spots between the leader and follower, none with developed penumbra (i). String these three categorizing letters together, and the McIntosh classification is "Dai."

Understand of course that a group classification does now and again change over the course of a day, when the group experiences shifts in appearance; just like a daily sunspot count it is a variable metric, only absolutely accurate at the time of the observation.

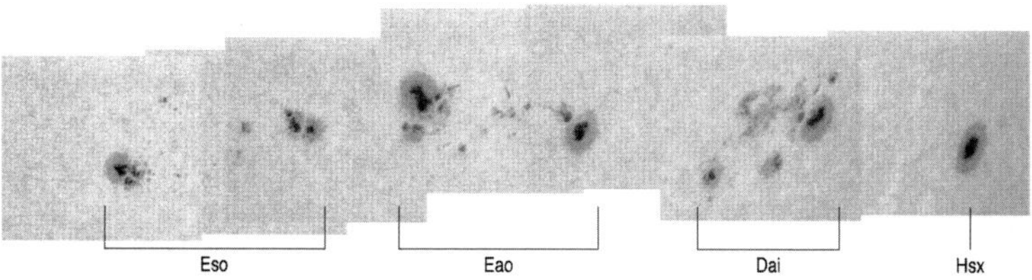

Fig. 2.18 Stretching across the face of the Sun on July 9, 2000 were four sunspot groups having differing McIntosh classifications (Courtesy of Art Whipple)

If direct observation is your mode, use a magnification suitable to determine the characteristics of the sunspot group. Usually no more than 50–100X is necessary, the seeing conditions are usually the limiting factor regarding magnification. No specific supplementary filters are necessary; however an orange or red filter will darken umbrae, making tiny spots between the leader and follower a bit easier. Individual spots, a unipolar concentration or any group lacking a penumbra will be accented too. Visually determining a spot diameter or a group's length takes practice. A reticule with a scale is useful when measuring dimensions directly, or a crosshair when timing a feature's drift in order to calculate heliographic dimensions.

Projection observers can accurately sketch a group's position and appearance on a report form and use separate templates to determine longitudinal length. Correction for limb foreshortening is accomplished with an angular measuring template.

Practice makes perfect in the classifying of sunspot groups. New observers will find it to their advantage to compare daily observations with those classifications published by professional astronomers. Mees Solar Observatory in Hawaii provides sunspot classifications daily on their web page. This activity will sharpen your skills, and develop an intimate comprehension of a sunspot's lifecycle.

If you are intent on observing an elusive white light flare, noting the McIntosh classification of groups on the visible disc of the Sun is essential to recognizing when and where to search (Fig. 2.18).

Observing White Light Flares

One of the most energetic yet least viewed events in our solar system is the white light flare (WLF). Solar flares are a manifestation resulting from the swift release of energy that has built up between opposing magnetic fields in an active region. A flare is generally best seen in the light of H-alpha, its outlet of greatest visible light emission. Flares go off several times per day during solar maximum while the solar minimum produces relatively few flaring events. Historically, it is noteworthy to mention that the first solar flare observed was a WLF by Richard Carrington, and confirmed by Richard Hodgson. The date was the 1st of September 1859.

The H-alpha flare (see section "Solar Flares" later in this chapter) is often seen beginning as a bright point or two, growing in size and brightness over a period of several minutes to a few hours. The rise to peak brightness in a solar flare is called the flash phase. To become a WLF, a solar flare must become so energetic and intense that its light "spills" out of the spectral regions of normal emission, and into the surrounding solar continuum. When that happens, the flare becomes visible at wavelengths outside H-alpha, which is the white light of the photosphere. This peak release of energy when the flare is glowing in white light is conspicuous for perhaps 5–10 min during the flash phase. Consequently, observing at the right time is *everything* with this type of observation. A white light flare appears as a single knot, or possibly two blobs located on either side of the neutral line or maybe within the spot's penumbral region. Expect an intensity approaching 1.5 I_{phot}. Although truly bright examples are rare, astronomers believe that a less intense variety of WLF can be observed more often.

The keys to spotting any WLF are deciding when and where to look, developing a methodical approach to the search, and optimizing the telescope to increase your chances of finding the elusive flare. Sunspot groups that have developed into McIntosh classifications of D, E, and F often produce flares. Chances increase for those groups that are sub-class ki or kc, though not limited to those classes; groups of Dai, Dso, and Hsx have, on occasion produced flares.

Begin by familiarizing yourself with the classes of groups currently on the Sun. Pay close attention to irregular and/or detached penumbra within a sunspot and areas of clustered spots between the leader and follower.

Table 2.2 Keys to locating a white light flare

1. Search McIntosh classes D, E, F, particularly those sub-classes of ki and kc
2. Set up an observing time when frequent checks can be made at the telescope
3. Observe in blue light inclusive of the G-band, where flares are known to go into emission

As was said, timing is critical and an occasional look at the Sun will hardly lead to success in spotting a white light flare! Most of us cannot monitor a possible flare producing sunspot group continually, but it is reasonable to take a few moments throughout a weekend morning or afternoon, such as when working outdoors or gardening, to scan the suspected sunspot group periodically for any WLF activity (Table 2.2).

Because only the brightest WLFs will be visible on a projection screen, use the direct viewing approach for your search. A prudent choice in the filtering system is the best way to optimize a telescope for observing a WLF. The idea with any optimization is to boost the contrast between the flare and surrounding photosphere, do this with a visual density objective filter principally transmitting in the *blue* spectral region. Mylar filters such as the Tuthill Solar Skreen fall into this category. You can take this a step further by combining a mild blue eyepiece filter with the safe mylar objective filter to intensify contrast of the flare a bit more.

You can also search for WLFs using video imaging methods by using a narrower band filter (10 nm or less) centered near 430.0 nm in the solar spectrum rather than the standard eyepiece filter. This filter will encompass the so-called G-band, which also goes into emission during a solar flare. This practice will dramatically raise your chance of catching a WLF, but the transmission characteristics of this narrow band filter may necessitate using a photographic density objective filter. The increased amount of light transmitted with the photo density filter is **not** safe for visual observing, therefore only photographic or video observations are acceptable.

When you spot a WLF, note to the nearest second if possible, the time of the beginning of the observation and the flare's disappearance, where on the Sun it occurred, its appearance, and a determination of relative brightness compared to the photosphere at specific times. Report these observations to the appropriate recording organization, such as the A.L.P.O. Solar Section or the B.A.A. Solar Division (Fig. 2.19).

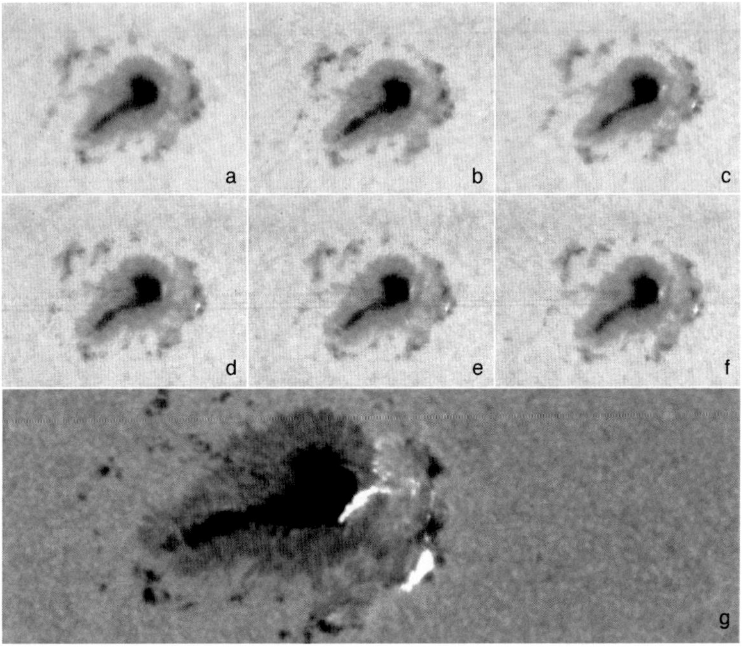

Fig. 2.19 Fortunate capture of a WLF. These images (**a** through **f**) were collected on 24 November 2000 by veteran observer Art Whipple. The photos illustrate the minimal time involved in a WLF's flash phase; only 1 min has transpired from (**a**) (15:08 UT) when the flare first became visible as a bright point until peak brightness is attained at (**f**) (15:09 UT). Image (**g**) is for comparison, obtained near peak brightness from outer space by the Transition Region and Coronal Explorer (TRACE) spacecraft, a mission of the Stanford-Lockheed Institute for Space Research and part of the NASA Small Explorer program (Courtesy of Art Whipple)

Lower Chromosphere

Directly above the photosphere is the atmospheric layer known as the chromosphere. During a total eclipse of the Sun the chromosphere (meaning sphere of color) is glimpsed as the pinkish-red ring encircling the solar limb. A nearly transparent gaseous region, the chromosphere is about

2,000 km thick (no more than 10 arc sec at the limb) and has an average temperature near 10,000 K.

Novice observers often wonder why on ordinary days they can't see the chromosphere through their standard "white light" telescope. The answer is quite simple: the light of the photosphere outshines the relatively feeble light of the chromosphere. In the past, other than during a total solar eclipse, chromospheric observing was accomplished with instruments that dispersed or broke apart the Sun's light into its various components. After separating the many wavelengths, the astronomer was able to view selected spectral regions where the chromosphere is brightest, *and* the photosphere becomes relatively opaque. These instruments, the spectroscope and spectrohelioscope, are powerful yet delicate tools that facilitated the first continuous studies of the chromospheric limb, and the first observations of the chromospheric disc.

Today's typical amateur astronomer rather than considering spectrographic techniques most often utilizes the add-on monochromatic filter, or a special solar telescope that passes only a specific wavelength of chromospheric light.

There are two wavelengths in the solar spectrum an amateur studying the chromosphere commonly makes use of: the Calcium-K (393.3 nm) or H-alpha (656.3 nm). When H-alpha or Ca-K features are viewed, light outside their wavelength is suppressed and remarkable details become visible. Filters are available for other specialized studies of the chromosphere including the Ca-H line, Hydrogen Beta, Sodium D, and Helium D3. These are dedicated products that optimize the viewing of select solar characteristics that may not be glimpsed in the more popular filters.

Ca-K filters make discernible lower chromosphere features which are affected by local magnetic fields. For instance, weak magnetic fields determine dark zones; stronger magnetic fields create brighter patches; however rather unexpectedly the strongest magnetic fields (i.e. sunspot umbra) look the darkest (Fig. 2.20).

Since the calcium spectral lines are broad, a filter with a bandwidth of 2.2 Å is suitable to isolate the K-line. The typical Ca-K filter will show a region approximately 500–2,000 km above the photosphere. This encompasses

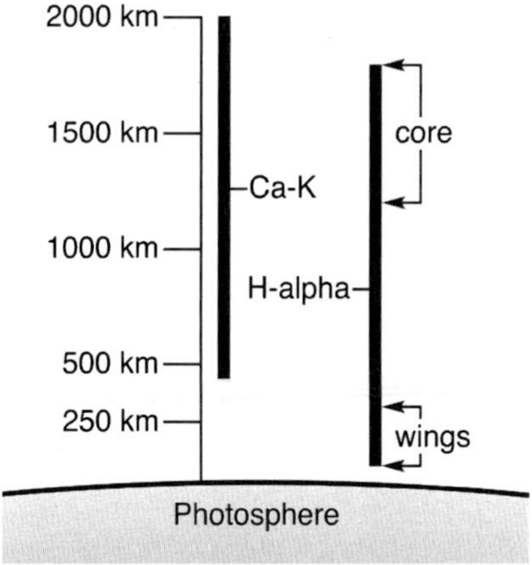

Fig. 2.20 The chromospheric heights visible through Ca-k and H-alpha filters (Data courtesy of Alexandra Hart)

the core height of the H-alpha line, so some Hα features will be visible, but for the most part they remain faint and ill-defined given that they emit weakly in the light of calcium (Fig. 2.21).

K-Grains

K-grains are observed in monochromatic Ca-K or Ca-H light and have a correlation to and similar appearance to continuum bright points. Discovered by Hale and Ellerman (1903–1904) they were originally called "minute calcium flocculi" with solar astronomers later adopting "K-grains" in the 1960s. As with bright points, K-grains require superb resolution due to their diminutive size. K-grains experience short (approximately 10 min) lifetimes in which they are seen to brighten and dim repeatedly while sometimes drifting away from their original location. K-grains tend to form solely in areas away from active regions within the interior cells of the chromospheric network.

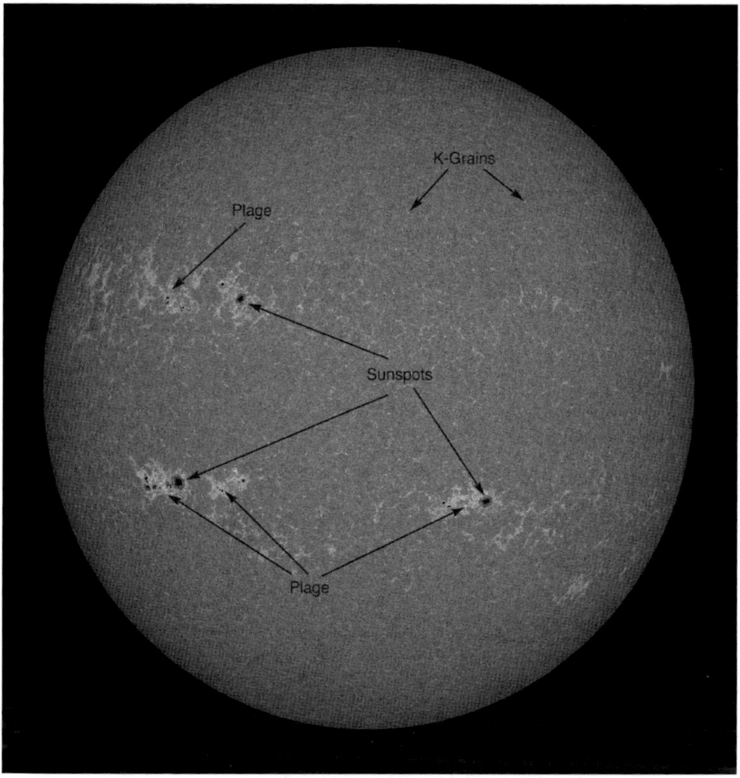

Fig. 2.21 The lower chromospheric Sun visible through a 393.3 nm Calcium-K filter on 30 June 2012. Indicated are the most obvious features: sunspots, plage, and K-grains (Courtesy of Christian Viladrich)

Chromospheric Network

The chromospheric network is a low contrast, fine mottled latticework or web-like structure that covers the Sun's face. In the quiet solar regions it is visible with slight difficulty in the light of Ca-K as a bright pattern, and with a bit greater difficulty (better in the wings) in H-alpha reversed, the web pattern then appearing dark. An improved visibility is noticed near active regions and plage.

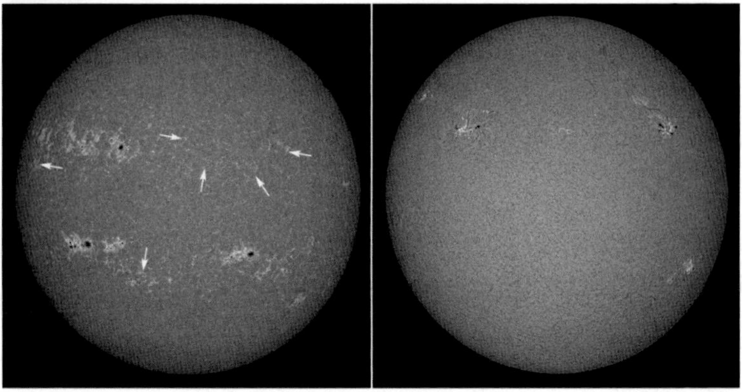

Fig. 2.22 The chromospheric network appears as the weak yet bright background pattern visible in the *left hand* Ca-K image, and *dark* in the *right hand* H-alpha image, spicules and fibrils limit the view of the network (Images courtesy of Christian Viladrich and Fabio Acquarone, respectively)

The network overlays the supergranular cells found in the photosphere. *Supergranules* are enormous versions of individual granules, each with a diameter about 30,000 km, although usually not easily seen if at all in white light. A continuously evolving feature, supergranules possess a lifetime approximating a day or so. Fluid motions within the supergranules carry magnetic field lines to the periphery of the supercells allowing the formation of the chromospheric network.

Coarse bright mottles appearing as miniscule patches in conjunction with bright point filigree are located along the boundaries of the supercells forming the network. From the mottles protrude larger dark appearing (on the disc) gas jets called *spicules* and another feature, the *fibrils* that follow the horizontal magnetic field lines sometimes found in the chromosphere. In the literature these two features are occasionally referred to as *dark mottles*. The spicules and fibrils provide an effective means of preventing the observer an unimpeded view of the underlying chromospheric network (Fig. 2.22).

Plage

The French word for beach is plage. *Plage* (pronounced PLA-juh) on the Sun refers to the patchy bright regions with enhanced temperature found most often near sunspots. Plage are visible predominantly in Ca-K, but also well in H-alpha light. Spicules avoid plage, but they do cluster at its edges.

It is believed that a relationship of some sort exists between the plage of the chromosphere and the faculae, a photospheric feature typically residing in a much smaller area. Faculae are explained by depression sidewall leakage and scattered light; plage on the other hand, indicate areas where underlying flux tubes are emerging. Plage embodies many tiny, bright features. It has been estimated that on average the area of a plage is 30 % filled with emerging magnetic flux tubes. The reconnection of magnetic fields is another occurrence associated with plage. *Reconnection* is the rearranging of broken magnetic field lines with newer emerging flux tubes. Since the relationship between faculae and plage is not completely known, this remains one of the mysteries for the professional astronomer to solve. From the amateur observer's standpoint, remember that these are two separate features, and should not be confused as one and the same (Fig. 2.23).

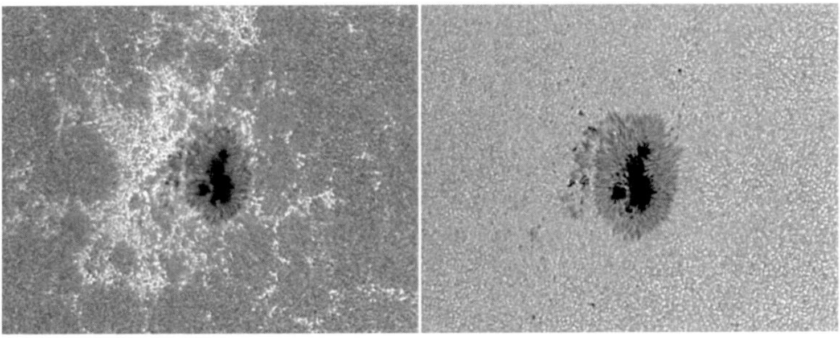

Fig. 2.23 Compare the two images of the same sunspot on the same day. The Ca-K filtered image on the *left* reveals a complex plage system about the spot. The *right hand image* was taken in the solar continuum through a green filter no plage is visible (Courtesy of Christian Viladrich)

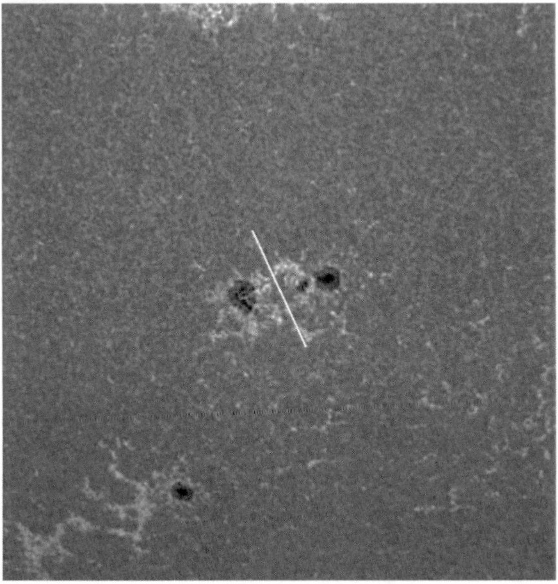

Fig. 2.24 The *white line* in this bipolar sunspot group symbolizes the area where magnetic polarity of the group reverses. The actual neutral line may not necessarily be a line segment but may follow an irregular path. Notice the plage has separated, creating a "plage corridor" extending on either side of the neutral line (Courtesy of Alexandra Hart)

A lower chromosphere feature, a plage can surround a sunspot as a cloud-like form with no particular consistency in brightness or shape; plage intensity disperses gradually as it reaches deeper into the less active regions of the Sun. Just as faculae precede sunspot development, small elliptical shaped plage may appear prior to sunspots. These early seen plage are customarily bright and may accompany arch filaments located within an emerging flux region (EFR). As the sunspot group grows, and if it is bipolar, the plage divides between the leading and following spots along the group's neutral line (a pathway where magnetic polarity reverses); this division is called the *plage corridor*. Early sunspot groups are noted for their sharp, thin corridor that becomes blurred with time, eventually disappearing and dissipating as the host group and attending plage decay (Fig. 2.24).

Upper Chromosphere

Those chromospheric features, excluding the aforementioned, are attributed to the mid to upper chromosphere, and are often viewed with a filter passing light at a wavelength of 656.3 nm, the H-alpha line.

The H-alpha filter is generally positioned to isolate the core of the spectral line (<1.0 Å) and show the mid-region of the chromosphere or about 1,500 km above the photosphere. Observing in the "wings" or off-band, up to about 2 Å in H-alpha, brings the view down much closer to the photosphere, somewhat less than 500 km from the surface.

Some mid to upper chromospheric features are visible in both the light of H-alpha and Ca-K, though in Ca-K rather fleetingly if at all. Features glow or are in emission at certain spectral wavelengths, affecting how bright they appear at other wavelengths. Consequently, prominences while bright in H-alpha are relatively faint when viewed in Ca-K (Fig. 2.25).

Spicule Patterns

Viewed in H-alpha at the limb of the Sun, a spicule appears bright because it is an emission feature seen before the cool, dark background of outer space. Spicules form the pinkish-red ring encircling the Sun that is seen during an eclipse, resembling what many early observers thought were "gas-jets". There is a tendency for spicules to merge together, making it exceedingly difficult but not impossible to recognize individual jets. They are seen to project from the limb at an angle near 70°–90° and often are larger at the polar regions of the Sun, protruding no more than 10 arc sec above the surface.

Spicules are hot gas, moving vertically at a speed approaching 50,000 km/h, and then abruptly ending their life in less than 10 min, by either falling back into the Sun or disappearing completely.

On the disc of the Sun, the spicules become absorption features (also called dark mottles) viewed before the brighter solar disc, creating interesting patterns given aesthetically pleasing names. Clusters of spicules are

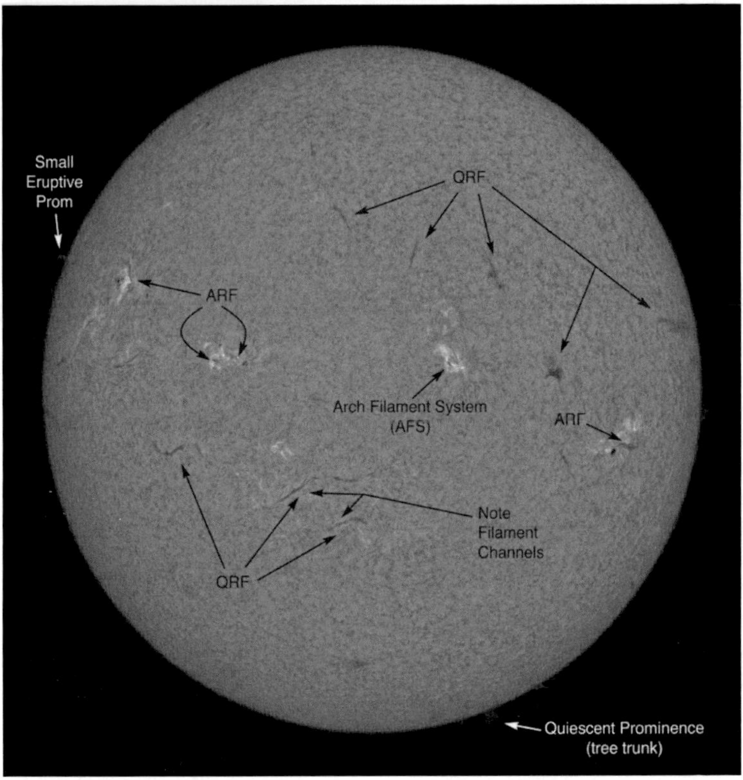

Fig. 2.25 Whole disc image with H-alpha features. A number of quiet region filaments (QRF) are spread over face of the Sun. Two located near the *lower center* have a *light appearing undertone* where it is evident that the spicules and fibrils are bent to provide the filament an opening, the filament channel. *Right of center* within the plage region is seen several members of an arch filament system (AFS), while to the *left of center* are located two active regions and several smaller active region filaments (ARF). The solar limb contains a variety of prominence features including a quiescent (tree trunk) prominence, and what appears to be a small erupting prominence (Courtesy of Alexandra Hart)

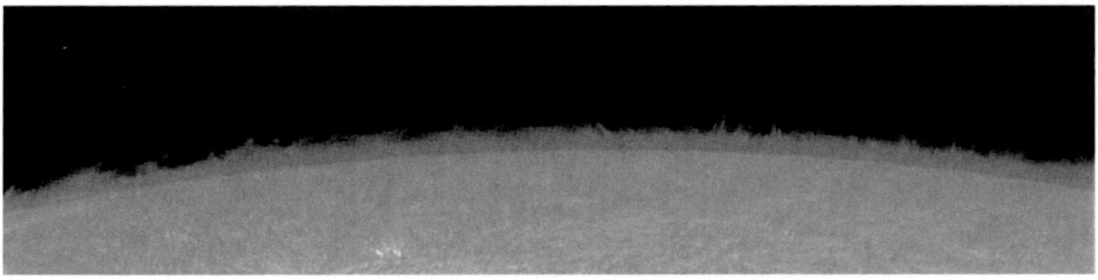

Fig. 2.26 Spicules on the limb of the Sun (Courtesy of Eric Roel)

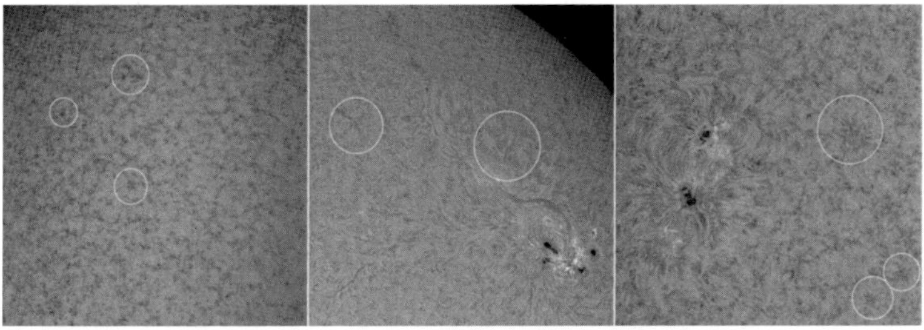

Fig. 2.27 Spicules on the solar disc. In the *left image* bushes can be seen *dotting* the image, three examples are *circled*. The *center picture* has two chains *circled*, the left most illustration lies beneath the filament. The significant difference between the bush and chain is in alignment. Bushes are clusters of spicules while chains are in columns and rows. *On the right* three rosettes are shown (All images courtesy of Fabio Acquarone)

referred to as a *bush*. Under good seeing conditions the fine spicules of a bush provide a striking multi-dimensional effect. If the spicules are aligned to form a column or row, they are known as a *chain*.

One of the more eye-catching shapes is that of the *rosette*, the radiating of spicules from a central point, like that of a flower petal. Rosettes form at the crossroad of two or three supercells of the chromospheric network. The radiating pattern is fashioned from bright mottles at the rosette's center, and alternating bright and dark mottles about the center (Figs. 2.26 and 2.27).

Fibrils

An active region containing sunspots with a strong magnetic field has influence over the neighboring spicules. Those located nearby become bent and enlarged, their projections following the local magnetic field lines. Visible in H-alpha light, this sometimes results in spectacular patterns of swirls, carpet-like formations, or whirlpool type effects.

These affected spicules are known as *fibrils*, dark mottles that follow the transverse or horizontal magnetic fields that run parallel to the solar surface. Individual fibrils can be 10,000–11,000 km in length, and display thicknesses of about 2,000 km.

When the thin gas in the upper chromosphere adheres to a local magnetic field, the finger-like fibrils are seen that indicate the location of lines of magnetism. Fibrils are known to unite regions of opposing polarity. Remember the science class experiment with the iron filings and the bar magnet. The same magnetic alignment principle is in effect on the Sun, often times visible in a bipolar sunspot group. Large sunspots are known to cause fibrils to diverge around a spot, giving the appearance of the thickening of a penumbral region. This effect is known as a *superpenumbra*. Exceptionally large groups can establish equally large regions of radial swirled fibrils, a condition that is designated a *vortex* (Fig. 2.28).

Emerging Flux Region

Earlier in this chapter the basic scheme of how sunspots are birthed, as seen in white light, was presented. That process began with the appearance of facula and pores. Some of the pores decay as others grow and develop into larger sunspots, and so on. There is a piece of the story, which the white light observer is not privileged to witness, but the monochromatic observer can view.

To more fully understand the development of an active region we must begin with the processes working deep in the Sun. Recall from Chap. 1 that the Sun generates a magnetic field between the radiative and convective zones in an area called the tachocline. The resultant lines of magnetism

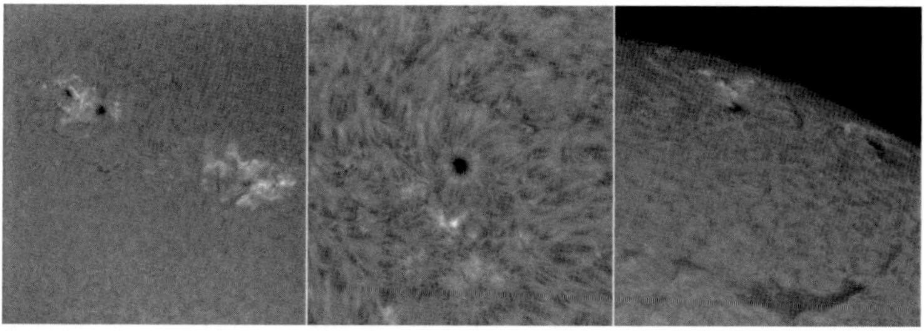

Fig. 2.28 Spicules near strong transverse magnetic lines bend and become fibrils. Fabio Acquarone obtained the far *left image* illustration showing the classic "iron filings" alignment found about the *upper left corner* bipolar sunspot group. In the *center* is a unipolar group, in which the fibrils radiate from the spot, giving the false impression of an enlarged penumbra or superpenumbra. *On the right* two sunspot groups appear at the solar limb with a swirling vortex of fibrils surrounding them (Center and right images courtesy of Alexandra Hart)

running from the Sun's north and south poles are wrapped around and around the Sun by differential rotation. Over time the vertical motion of plasma in the convection zone causes a kinking or tangling of the elongated magnetic field lines. The tangled lines develop regions of increased magnetic strength in the convection zone known as magnetic flux tubes. A process called buoyancy allows a flux tube to rise to the surface and poke through the photosphere. Basically, that is the process of magnetic flux emergence.

As the magnetic flux tubes initially break the surface, areas known as *ephemeral regions* develop which are not nearly as bright as plage, and contain no pores. These areas are known to also occur outside the sunspot zone, but most are formed within it.

To the monochromatic observer the initial beginning of an active region, known as the *emerging flux region* (EFR), is the early development of plage. Also seen in Ca-K, tiny flux tubes often less than 600 km in diameter poke through the solar surface allow a peek into the solar interior, hence the reason they appear bright. Some small tubes will merge together to form larger, more intense structures which begin to stifle convection forming

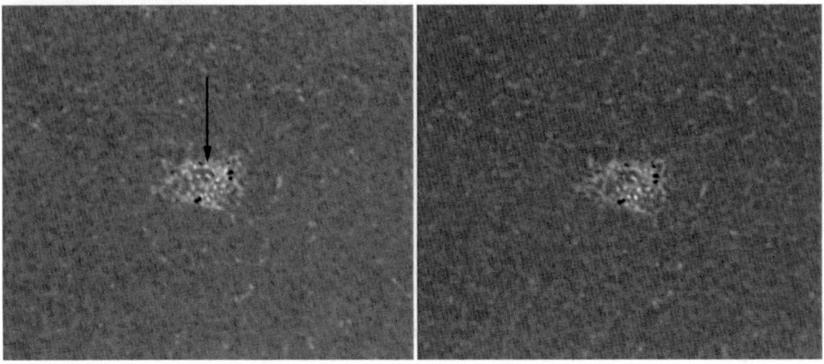

Fig. 2.29 Emerging Flux Region prior to being designated AR1130 by Alexandra Hart from 28 November 2010. The *arrow* indicates several "bright points" which may be Ellerman bombs. Image on the *left* is from 12:37 UT, *on the right* at 13:02 UT (Images courtesy of Alexandra Hart)

pores, which then become visible to the white light observer. Several other monochromatic features are known to accompany an emerging flux region, including the arch filament system and Ellerman bombs (Fig. 2.29).

Arch Filament

During the emergence of flux, as an active region grows, a unique type of fibril may be observed in H-alpha, the *arch filament system*. An arch filament system (AFS) when observed from above in early plage is composed of dark, thick threads that always overlay the bisecting neutral line of the active region. Their form may materialize as straight lines or include slight curvatures. Because an arch filament is low in the chromosphere, observing it at the limb can be difficult depending on its height, but not impossible. On the limb an arch filament is visible as an emission feature, bright against the dark sky background.

The AFS, because it follows the magnetic field lines of opposing polarity is frequently an arch-shaped feature. Height can vary from system to system with an observed variability of 4,000–15,000 km. The filaments tend to last around 30 min, their transformation seemingly tied to visible changes

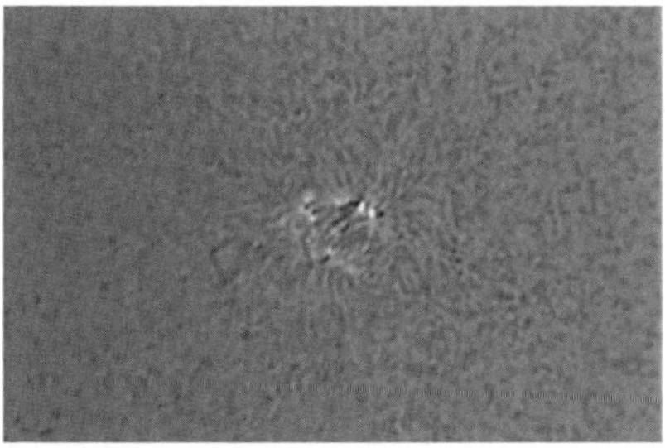

Fig. 2.30 Arch filaments found in this emerging flux region designated AR1130 from 28 November 2010 are seen as the *thick, dark threads* crossing from *lower left* to *upper right* over the numerous bright points (Image courtesy of Alexandra Hart)

within the emerging flux region. Three days is the accepted period for a developing sunspot group to harbor an AFS.

It is theorized that arch filaments are caused by the constriction of plasma consequential from the new emerging flux, which as the plasma becomes denser creates a flow downward at both ends of the arch due to the influence of gravity. What we usually see through the spectroscope in the AFS is upward flow (Doppler shifted blue) near the peak of the arch, and downward flow (Doppler shifted red) at the arch's feet or foot points, which will be anchored in the active region, at the flux tube points of opposite polarities. As a sunspot group matures you will find the AFS is transformed into a *field transition arch* (FTA), similar to the AFS, but more stable and possessing weaker appearing thin, gray, fibril-like features also anchored in the magnetic bi-poles (Fig. 2.30).

Ellerman Bomb

Another typical feature associated with the emerging flux region and mature active regions are the bright points, having a diameter of 3 arc sec or less, known as Ellerman Bombs or in spectrographic circles, *moustaches*. Ferdinand Ellerman, a contemporary observer of George E. Hale who likened the brightenings to solar hydrogen bombs, is their namesake.

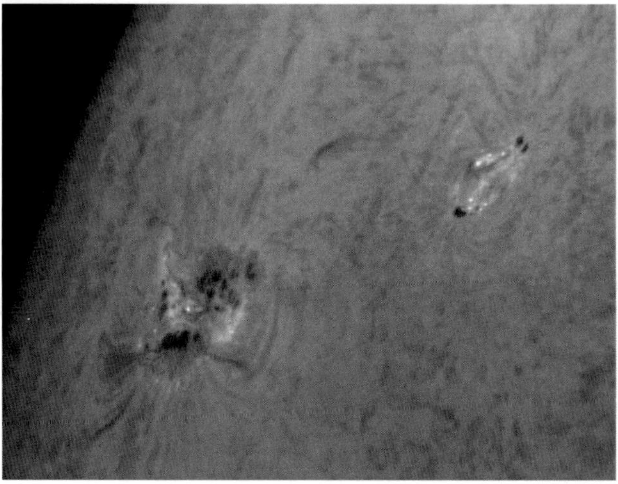

Fig. 2.31 Ellerman bombs (*small dots*) visible around the perimeter of AR1429 and the nearby emerging flux region on 4 March 2012. Image obtained −0.5 Å off-band from H-alpha (Courtesy of Jim Ferreira)

When observed spectrographically the appearance of the bright point is that of a strong absorption in the centerline of H-alpha and a bold emission in the wings of the line, hence the name moustaches.

Ellerman bombs occur low in the chromosphere, 600–1,100 km above the surface, and therefore appear best in the wings of the H-alpha line. Bombs last in the range of 5–10 min before disappearing and have been known to pop back into view after a half hour or so. Look for Ellerman bombs within the interior of a new emerging flux region, around the edges of sunspots, or at the base of spicules and active filaments. Ellerman bombs figure to be the result of magnetic reconnections in the chromosphere (Figs. 2.31 and 2.32).

Prominences and Filaments

Other than the solar flare, a prominence is the most stunning event to be seen on the Sun. The prominence (abbreviated in amateur circles as the prom) is a cloud of gas suspended above the solar surface. The cloud may

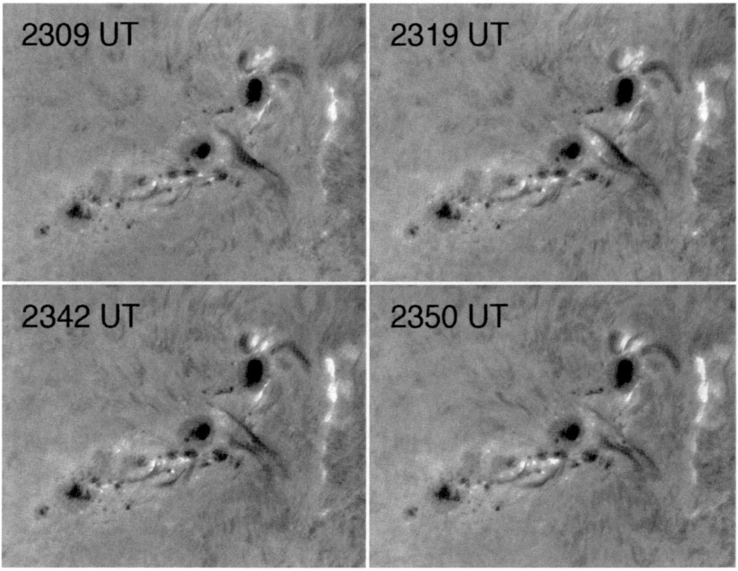

Fig. 2.32 Changing appearance of Ellerman bombs in AR1515 on 3 July 2012 between 2309 UT and 2350 UT. Jim Ferreira imaged this event −0.5 Å off-band from H-alpha (Courtesy of Jim Ferreira)

condense out of the lower corona, be thrown up by an ejection from below, or define a sheared neutral line by gas captured amongst the twisted bundles of magnetic field set up above the photosphere. The novice and at times the general public often make the mistaken assumption that a prominence and a solar flare are one in the same feature. The two can and do have a link, but they are totally different features; a prom being as defined above, while a solar flare is the swift release of energy accumulated within the magnetic field of an active region.

Typically a prominence has a relatively cool temperature near 10,000 K and a density hundreds of times that found in the immediate chromosphere and corona. Dimensions of a prom can vary, but a height of 40,000 km and a length of hundreds of thousands of kilometers are easily attainable. The wise solar observer always keeps in mind this "tongue in cheek" rule regarding a prom: "when it comes to a prominence, almost anything is possible".

Viewed ideally in H-alpha and at the solar limb, prominences are beautifully profiled features seen in emission (bright) against a dark background sky. Under these conditions proms can be watched with a filter having a bandwidth of 1–2 Å. A wider bandwidth filter up to about 10 Å (consequently passing some continuum light) will also show limb prominences, but an occulting cone or disc is necessary to safely block the blindingly radiant light of the photosphere. To view a prominence upon the solar disc requires an H-alpha filter with a bandwidth less than 1.0 Å, where unless the prom is especially energetic, it will appear as an absorption feature, a gray to black ribbon-like structure known as a *filament*. Prominences and filaments are the same feature, they just are observed on the Sun at different locations.

Filaments associated with a sunspot group tend to be narrow and dark, sometimes winding among the group; the longer, gray, thick varieties are found in the quiet regions of the Sun, changing their appearance over time ever so slowly. A prominence associated with an active region is identified as an *active region filament* or ARF. A prominence appearing in a quiet spot on the Sun is called, a *quiet region filament* or QRF.

Proms/filaments are not limited to the sunspot zones either; a filament can appear in the higher solar latitudes, practically anywhere on the Sun. In fact, located about 10° outside the sunspot zones is a "prominence zone", which during the 11-year solar cycle migrates toward the solar equator just as the sunspots do.

Because plasma clings to magnetic field lines, and magnetism supports prominences, a prom will outline the shape of its host field. Note that as the magnetic field evolves, the prom's shape will change. This characteristic provides one channel for mapping magnetic fields as they occur on the Sun.

There exists a significant connection among a prominence, a solar flare, and an active region. A prom/filament is likely to develop where conditions result in the elevation of dense gas above the solar surface. One possible condition comes about as a solar flare manipulates material through thermal instability: resulting in cool gas condensing in the lower corona, and the subsequent condensation raining down onto the Sun's surface. Another connection related to all proms and filaments is the shouldering effect of gas by the magnetic fields produced by flux tubes located inside and outside active regions that poke through the photosphere.

Thirdly a solar flare, one of the most violent energy events in our solar system can disturb an existing prom/filament from the Sun. In most cases the ejected or erupted material falls back to the Sun; however a particularly spectacular eruption can lead to the ejection of tons of material from the corona, a so-called *coronal mass ejection* or CME. The CME if directed toward Earth frequently results in nightly displays of the aurora, and disruption of the near-Earth space environment. These three circumstances are found within photospheric active regions.

Through the years various schemes intended to classify and organize prominences have been proposed. Factors determining classification schemes have included: the physical appearance of the body (morphology), the prom's relationship to sunspots, motions of the prom, a feature's spectroscopic characteristics, or the prom's place of origin – the chromosphere or corona. The reality unfortunately is that no single scheme can easily describe *all* the possible forms a prominence assumes. Donald H. Menzel, John W. Evans and F. Shirley Jones in a system that will be presented later in this section did make an interesting attempt at prominence categorizing in the 1950s. The present discussion however is not focused on classifying prominences, but rather only describing the forms that a prominence assumes. Therefore, at this point only two divisions of proms shall be considered, those that to some degree are stable, and those that are not and have a bursting nature.

Early chromospheric observers, such as Fr. Secchi, using a wide slit on their spectroscope recognized two contrasting groups after having studied a large number of prominences. These groups are: *quiescent* (meaning quiet/static) and *eruptive* (meaning active/moving). For the most part a quiescent prominence behaves in a calm manner, changing appearance only slightly with time. One could say a quiescent remains somewhat sedate. The quiescent prom, identified with the previously mentioned QRF, often originates in the corona with matter **descending** toward the photosphere. On the other hand an eruptive prominence many times originates in the chromosphere with its matter ejecting or **ascending** toward the corona. Eruptive proms appear agitated, and by and large are associated with active regions, hence the acronym ARF applies.

Since the distance to the Sun is great, and even though some prominences are large, prominence morphology in real time does appear gradual. However, don't think that an eruptive prominence's development necessarily proceeds at a turtle's pace; dramatic changes can be seen within a minute or less! Eruptions may easily exceed 200 km/s. A quiescent prom does not always remain quiet either; it can become unexpectedly disturbed and erupt into space. And because an ARF or QRF may have erupted, don't think the show is over; the feature may reappear minutes or 1–7 days later in the same solar region.

As stated earlier, these features are best observed in H-alpha, where they are in brightest emission. However, occasional viewing off-band may be essential if an eruptive feature has been Doppler shifted. While it is possible to observe prominences and filaments in the light of calcium (Ca-K), the view most times is quite dim, and ill-defined.

Following you will find descriptive material that outlines some of the more commonly documented forms, beginning with the descending quiescent type and progressing to the ascending eruptive variety. It would take an entire book devoted to this one topic to do complete justice, so don't be surprised if during your own observing sessions you find an example or two that fail to fit within these familiar prominence types. Remember the rule of thumb regarding proms and filaments: that anything seems possible.

Descending Prominences

Many start the life process as a small active region prominence located perhaps at the perimeter of an AR where the prom connects to an outside source of opposing magnetic polarity (another active region or a new emerging source of magnetic flux, called a *dipole*) or positioned along the shearing neutral line of an AR. Many times a filament is found winding its way through a sunspot group, or a quiet region of the Sun where local fibrils tend to bend themselves into a pathway or opening parallel to the filament. This magnetically neutral region called a *filament channel* is situated below the filament and appears lighter, the result of restrained spicules and fibrils that are outlining the chromospheric network.

Initial development of a typical quiescent prominence may take 24 or more hours. The parenting active region, if one exists will eventually decay,

Fig. 2.33 The drifting of filaments to the polar regions of the Sun forms a polar crown. In this instance Thomas Ashcraft captured a string of filaments in the northern zone on 16 May 2010 (Courtesy of Thomas Ashcraft)

perhaps leaving the filament to broaden and lengthen into an enormous quiescent QRF, drifting gradually from the sunspot zone toward the poles.

Quiescent proms originating in the sunspot zones tend to have a shorter lifespan than those found in the higher polar latitudes, two solar rotations compared to about five, respectively. Differential rotation of the Sun causes the lengthening of filaments as they slowly drift toward the solar poles. Some of these drifting filaments make it to the upper latitudes, joining other filaments already there to form a so-called, *polar crown*. This unusual feature ordinarily occurs at about ±70° heliographic latitude (Fig. 2.33).

The quiescent prominence exits by slowly disintegrating, flowing down to the surface, or by becoming *activated* and erupting. Activated means that the normally static quiescent is undergoing unexpected change, and is experiencing increased movement. A quiescent prominence/filament under these conditions is termed an *active quiescent*. Visible disturbances resulting from activation include: physical growth of the prominence, a change in brightness (darker for a filament and brighter for a limb feature), or a disorderly shifting of material within the feature.

Nearly all prominences achieving heights in excess of 50,000 km (approximately 65 arc sec) are known to habitually erupt within a day or two. Passing shock waves (see Moreton Waves) emanating from a solar flare can cause a rapid waving effect to the prom/filament that precipitates activation. And lastly, a quiescent can become activated for simply no apparent reason.

Material flows are typically descending, flowing from the prominence to the surface region. Should flow appear to be between two or more quiescent structures, these features are said to be *interactive*. The nearly horizontal flow of knots and streamers between individual components is infrequently detectable, however when it is, it may include interaction with a form of celebrated prominence renowned as a *tornado* (see following pages).

The *hedgerow* is the representative descending quiescent prominence. The name is derived from the obvious similarity in appearance to the rows of trees and hedges left along the edge of a farmer's field, serving as a windbreak. The grouping can be composed of a number of individual proms (the row in hedgerow), each having a slightly differing form. Or it may be one large prom, a gaseous composite structure made of a vertical palisade of "threads and fibers", each thread having a diameter of less than 300 km. Time-lapse movies clearly show considerable internal motion as hedgerow gases stream down the threads and into the lower chromosphere. It is not clear how the gas in the hedgerow is replaced, but it must be or else the prom would exhaust its supply long before most disappear (Fig. 2.34).

The ends of the hedgerow are often magnetically anchored to the solar surface. Normally a stable, quiet feature, when one end of a hedgerow breaks free from below, this is understood to be a sign that the prominence

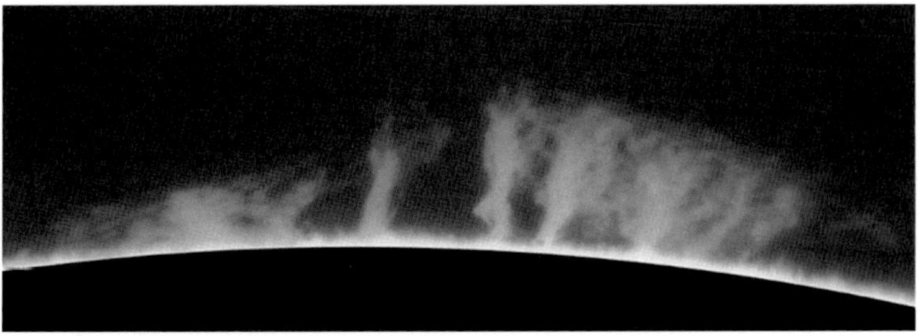

Fig. 2.34 A very typical hedgerow formation on the solar limb composed of a field of individual trees (Courtesy of Alexandra Hart)

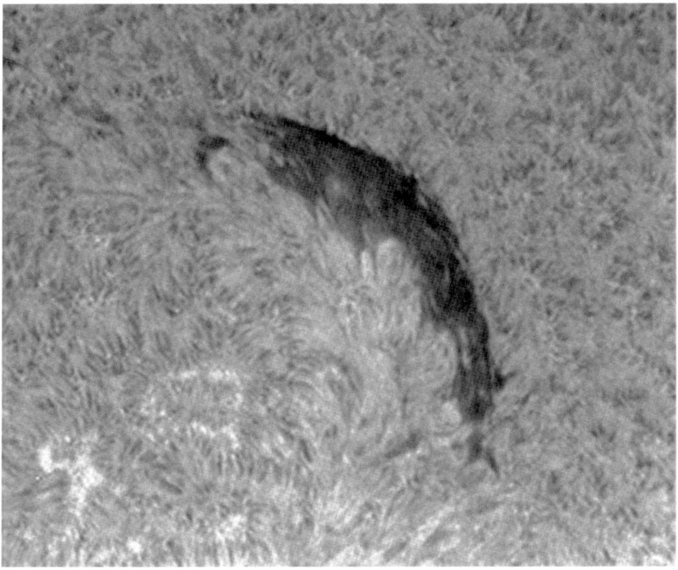

Fig. 2.35 Here is a quiescent filament located on the disc of the Sun showing a smooth *topside* and a *lower scalloped side* (Courtesy of Gordon Garcia)

is about to erupt. Viewed at a steep angle from Earth, as in Fig. 2.35, the filament sometimes presents a smooth edge on the upper side while the lower inside edge appears scalloped. These scalloped points are identified as *barbs*. The lower side projections often extend to the area where an active region magnetic field reverses polarity, the neutral line. The smoother upper side is less ragged because of its interaction with the supporting, and also constricting magnetic field.

Another feature of a quiescent prominence, particularly the hedgerow, includes so-called ascending *bubbles* and *plumes*. Contrary to the normal down flow occurring in the quiescent prom, dark convex-shaped bubbles are seen to develop unexpectedly below the base of the prom, but situated above the chromosphere's spicules. These bubbles subsequently initiate a small scale, turbulent up flow through the prominence that is dubbed, a plume. A bubble can spawn a number of plumes in its several hour lifetime, new plumes appearing approximately every 5–10 min. A superb illustration of the effect is found in the Ferriera images found in Fig. 2.36.

Other Descending Forms

Numerous other forms of the quiescent prominence exist; common nicknames describing their appearance include, but are not limited to the haystack, tree trunk, mound, and coronal cloud, floating arch, double arch, coronal rain and tornado.

Prominence loops are particularly beautiful limb structures that fall into three different categories. Following the outburst of an intense flare, the solar material ejected into the corona sometimes begins to cool and condense into a motionless "dot" above the sunspot active region. Intensity of the dot increases as it expands downward toward the AR outlining the first type, a so-called *post flare loop*, with material flowing back onto the underlying surface at nearly 10 km/s. Sometimes the material flow is down both sides of a loop, and at times it can be seen rising up one side and flowing down the other. Post flare loops often have a lack of uniformity to their brightness, the upper loop region routinely appearing brighter than the arches.

If the loop is incomplete and open near the top a second type of post flare loop develops, this is *coronal rain*, consisting of faint streams and knots of gas pouring back from the gap intermittently through the chromosphere. Many times a flare loop system associated with a strong magnetic field will consist of a grouping or cluster of numerous thin loops (approximately 20)

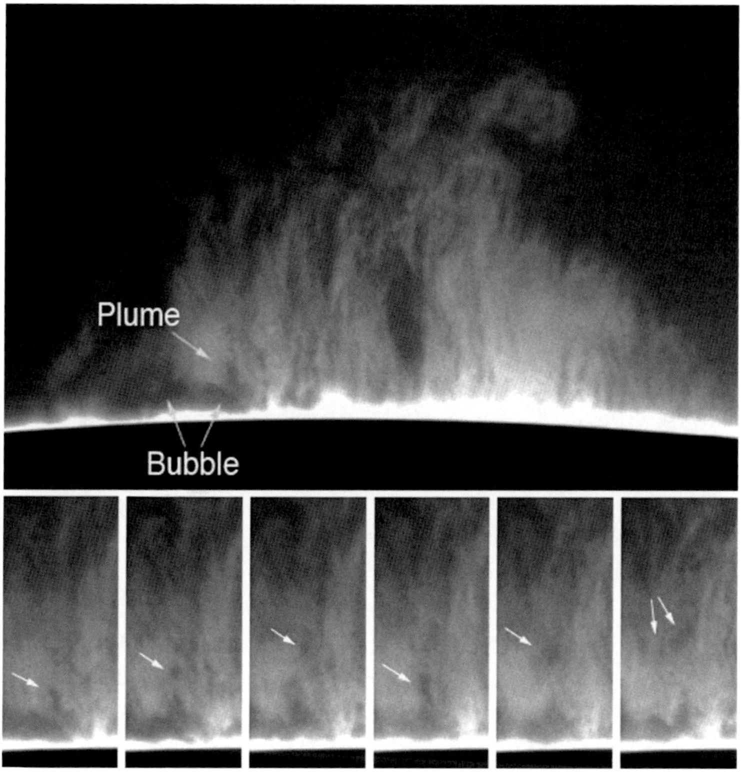

Fig. 2.36 Bubbles and plumes in a quiescent prominence. Bubbles of hot plasma develop instabilities, producing an up flow or plume of hot plasma that rises at nearly 20 km/s. Images extracted from a 42-min H-alpha movie on 28 August 2011 (Courtesy of Jim Ferriera)

called an *arcade*. Arcades are strikingly similar in appearance to a child's "slinky" spring toy with the arch foot points located near the active region's magnetic poles. This is to say, a loop crosses the active region's neutral line just as does an arch filament system. Loops are not long-lived features, several hours of activity is the norm.

The third type of loop is the *flaring arch*, similar in appearance to a post flare loop; the flaring arch develops as a flare expels the flowing material directly into a magnetic loop rather than condensing it from coronal material.

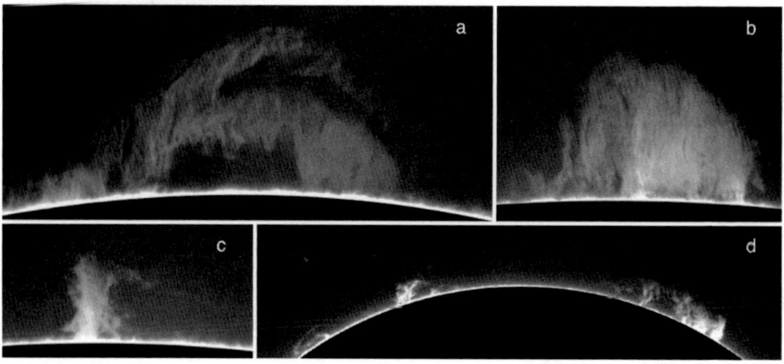

Fig. 2.37 Collection of descending quiescent proms. Figure (**a**) is a classic floating arch, (**b**) would be called a haystack or maybe a curtain depending on your point of view. If you look closely perhaps you can see the profile of a *facing-left elephant*, standing on the Sun. Another classic, the *tree trunk* is captured in figure (**c**). The wide field view of (**d**) contains a quiescent formation on the *right* and several eruptive proms on the *left* (Figure (**a**) courtesy of Alexandra Hart; Jim Ferriera provided the balance)

The *coronal cloud* is an ill-defined patch or blob, dissimilar from the dot of post flare loops, suspended in the corona with material streaming from it to the surface below. A good example can be found with Fig. 2.38a, nicknamed here because of its shape, the kite.

The *mound* and *pyramid* forms are similar in appearance other than the rounded peak of the mound and the sharper summit of the pyramid. Both are low bodies in which the near chromosphere width of the feature should exceed the height. The straight or curved *pillar* can fool you into thinking an eruption is happening. If the feature is long-lived and relatively motionless, the possibility of the pillar being an edge-on view of another QRF (i.e. floating arch, Fig. 2.37a) is good.

One of the more unusual prominence formations is the *tornado*. Solar astronomer Charles A. Young in 1910 referred to this prom as a "cyclone or whirling waterspout capped with a great cloud." The form is difficult to distinguish because a tornado is ordinary quite small, and its rotating nature is not immediately evident to the casual observer. Since some column-like prominence types do not rotate (i.e. pillar), spiral or helical motion of the prominence must be observed for it to be classified a

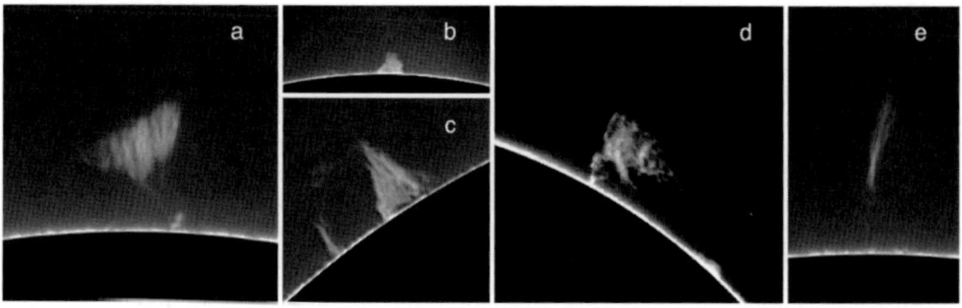

Fig. 2.38 A few unusual and not so unusual prominences. In figure (**a**) notice the *kite shape* and the apparent streaming of material from the corona. Closer inspection reveals two faint streamers extending from the "detached" coronal cloud to the tiny curved pillar feature below. A slightly deformed mound was captured in figure (**b**); often these become activated and erupt. Two formations appear in (**c**), on the *left* is a straight pillar while to the *right* is a pyramid. The pillar if turned 90° could possibly be another pyramid or maybe a mound. An unusual tornado was imaged in (**d**), with rotation spotted between several photos in the stubby "Y" shaped column. The smokestack emitted from the *top* appears "windswept". Figure (**e**) might be an unconfirmed tornado or just an anomaly of a prominence. In the original photograph two faint streamers (as in the kite) appear forming a column that extends vertically from the lower chromosphere to the seemingly detached material above (Figures (**a**) and (**e**) courtesy of Jim Ferreira, (**b**) and "(**d**) courtesy of Jamey Jenkins and (**c**) courtesy of Alexandra Hart)

tornado. The first confirmed tornado was spotted on 29 August 1869 by an observer named, Zollner. Color drawings done about an hour apart clearly indicate the twisting helical motion of this prominence, making it the first cyclone observable on the Sun (Fig. 2.38).

A tornado prominence comes in two varieties. The more common form has a column or pillar with a stream of matter flowing from its peak. Sometimes this stream is bent over and nearly reaches back to the chromosphere. The second form, seen less frequently, has a skeletal-type columnar structure, constructed of single, helically bound streamers with no matter emerging from the top. An increase in a tornado's rotational speed may result in its eventual obliteration, a rupture reminiscent of an Earthly whirling "dust devil". They have been known to even abruptly rise into the corona and fade away (Figs. 2.39 and 2.40).

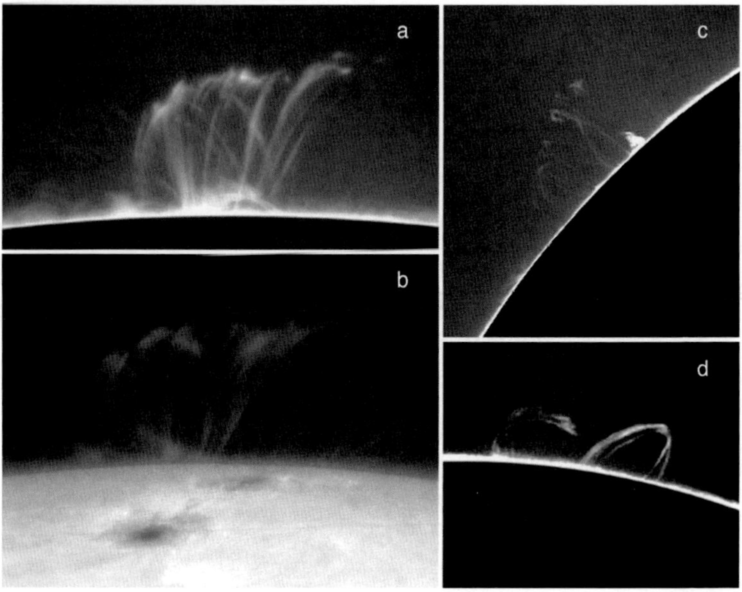

Fig. 2.39 Loops are one of the most spectacular sights in our solar system. The *left* figures (**a**) and (**b**) are of the same event from 22 September 2011. The height above the limb measures 70,000 km as (**b**) shows the numerous loops (arcade) emanating from foot points of opposing magnetic poles of the sunspot group. In (**c**) faint post flare loops and coronal rain appears above the sunspot group rotating around the east limb of the Sun. The streamers of rain are immediately to the *left* of the bright eruption occurring in the chromosphere. Figure (**d**) illustrates a classic loop and an unconnected loop. Notice the variation in brightness within each. The unconnected loop is likely to complete the connection as more material descends to the feet ((**a**) and (**b**) courtesy of Jim Ferreira, (**c**) and (**d**) courtesy of Jamey Jenkins)

Fig. 2.40 (continued) Look specifically for downward movement of the streamers and the bright knot (bomb) about to splash in the lower chromosphere. The event occurred on 31 October 2010, top frame at 1655 UT and the last frame at 1701 UT (Courtesy of Jim Ferreira)

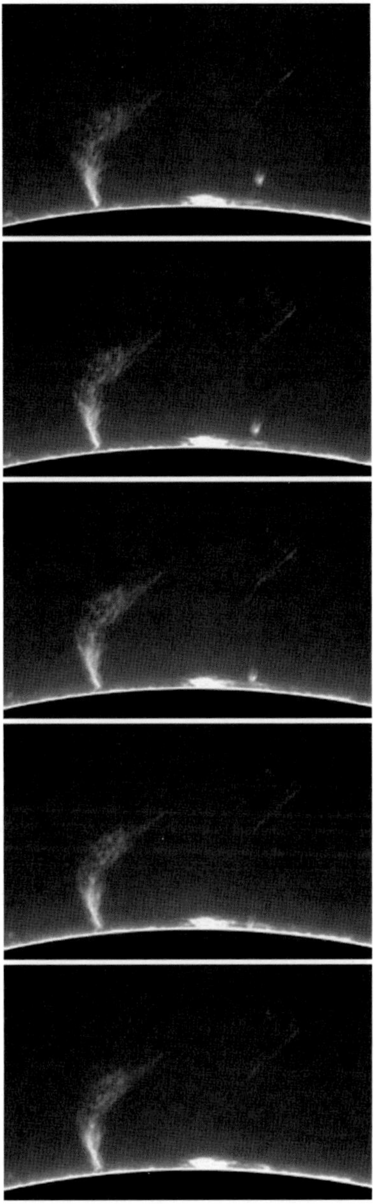

Fig. 2.40 This is a unique series of images (*top* to *bottom*) looking at the west solar limb by Jim Ferreira. He has photographed coronal rain and what Jim has dubbed a "coronal bomb". The descending material is falling from the *upper right* toward the bright flaring region near the center.

Ascending Prominences

When a prominence/filament erupts, it does so away from the Sun. An eruption can be a slow fading thing as with a large QRF, or a violent forceful projection having a velocity in excess of 2,000 km/s, as observed during an event called a spray. Regardless of the event, a prominence/filament as it ascends occasionally disperses material outside the gravitation influence of the Sun; even then an unpredictable amount every so often descends back to the chromosphere.

A filament may display several markers of an impending eruption. For instance, a QRF or ARF may exhibit an unstable motion, a cascade of material flowing through the filament or both. Moreover it would not be unusual to perceive an increase in brightness below an erupting filament. When a stable prominence reaches a height greater than 50,000 km, this is a sure sign of it becoming *activated* or triggered, and later erupting. Other visible signs of a possible eruption include an increase in the size of the filament or a change in intensity; that is a filament on the disc becoming darker and if observed on the limb substantially brighter. The activation process may last for an hour or so and in some cases activation unexpectedly dissipates as the prominence returns to a normal state.

An active region filament generally erupts swiftly, while a quiet region filament more often than not erupts slowly. The difference is found in the strength of the surrounding magnetic fields. Magnetic fields present about an ARF are much stronger than those found in the quiet Sun.

What is the root cause of an eruption? Outside disturbances intruding upon a prominence cause its eruption. The active region filament is often disturbed by the huge amounts of thermal and particle energies released during a solar flare or other magnetic disruption.

Activation follows when processes presented in a variety of differing scientific models, release large amounts of magnetic energy. One possible explanation says simply that as new magnetic flux emerges near older magnetic flux, a reconnection or realignment of the magnetic fields occurs. During this realignment process energy may be released. How the magnetic energy is converted to kinetic energy (explosive gas pressure) is unknown, and remains one of the mysteries of solar physics to be solved.

Quiet region filaments sometimes erupt for no apparent reason, but one known cause of activation is the passing of a shockwave called the Moreton wave. Initiated by a massive solar flare the Moreton wave will as it passes, lift a filament so that it oscillates several times precipitating activation while it causes the filament to become Doppler shifted in and out of visibility. A filament in this situation is termed a *winking filament*.

Disparition Brusque

A French expression used in filament studies, *disparition brusque* (DB) means literally to lift off or the "rapid disappearance" of a filament. Mature quiescent proms become activated, perhaps by a Moreton wave or a nearby emerging flux region and begin a slow rise of several hundred kilometers/second into the corona. The rise of the disparition brusque normally lasts less than a few hours. Sometimes the cause of the activation is unknown (see section "Hyder Flare"). The filament may simply fade away, or gradually break up and disappear as it rises higher and higher.

In certain cases involving large filaments, a DB precipitates the expulsion of several billion tons of matter from the Sun. This coronal mass ejection or CME may result in the ejected particles being swept to our vicinity in space by the solar wind, prompting electrical havoc and initiating beautiful auroras on our planet.

Surge

A surge is a controlled or collimated, relatively continuous expulsion of material, blown away by an underlying low-level compact, or simple-loop solar flare. It is a post-flare event possibly compelled by pressure or gravitational effects. A surge is collimated because it is bound by the magnetic field of the AR's primary sunspot. A surge will often burst outward at any angle from vertical to nearly horizontal attaining a possible elevation of several hundred thousand kilometers, lose inertia, and perchance fall back onto itself with a great splashing effect. The velocity of a typical surge is less than 200 km/s, less than the escape velocity of the Sun, which at the photosphere is about 600 km/s. A period of activity lasting perhaps 10–20 min is expected for small surges, and about 60 min for the largest examples. The event may reappear hourly in about the same location.

A surge viewed on the disc of the Sun is, by and large, moving in the direction of the Earth and unless especially energetic will appear dark (absorption), growing elongated (comet-like appearance) from the direction of the flare, and Doppler shifted toward the blue wing of the H-alpha line. The Doppler Effect results from the compressing or stretching of light waves because the object is rapidly moving toward or away from the observer. To the observer of a Doppler shifted eruption, the surge will begin to fade, gradually becoming transparent until it has disappeared completely. By tuning the center wavelength of an H-alpha filter toward the blue wing, the filament will reappear unless it has been completely obliterated.

When detectable at the Sun's limb, a surge appears as a bright jet or spike in emission, which grows swiftly, and then either falls back into the chromosphere along established magnetic field lines, or breaks apart and disappears completely.

The *puff*, also called a *smokestack*, is a smaller variant of the classic surge, with an appearance suggestive of its name. Visible as a blue shifted feature with a lifetime of several minutes, they are seen in an active region as well as in an emerging flux region. A puff/smokestack appears as an individual knot of material ejected at surge velocity disappearing completely, or like the surge falling back along the lines of its confining magnetic field (Fig. 2.41).

Spray

While a surge is a continuous burst of collimated material, the spray is a single ejection of filament material in an uncontrolled pattern, with matter going in various directions. Again the volatile nature of a solar flare is the progenitor, but in this case the flare is much larger. The velocity of many sprays exceeds 200 km/s and have been known to go beyond 2,000 km/s, easily attaining the Sun's gravitational escape velocity.

Two categories of spray are recognized: the *flare spray*, resulting in the throwing out of plasma due to the violent effect of a solar flare, and the *prominence spray* which is the eruption of an overlying filament above the neutral line.

The flare spray is visually impressive with matter scattering over a large area of the Sun. A prominence spray may begin with the lofting of a

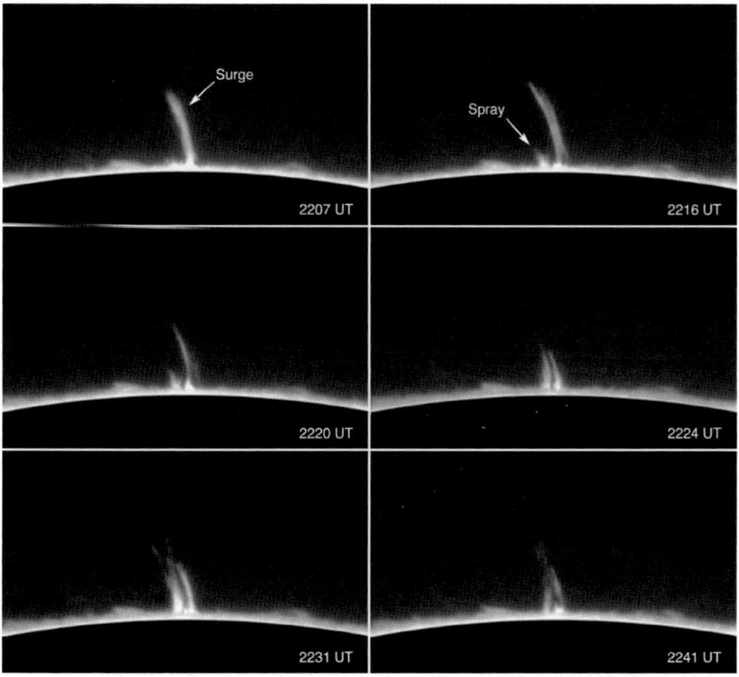

Fig. 2.41 Two events are happening in this series of images obtained 6 August 2011. At 2207 UT the *tall pillar* (surge) has erupted, successive images capture the ejected matter falling back along the lines of magnetic force near their place of origin. At 2216 UT a spray has erupted to the *left* of the surge, expelling matter in several directions (Courtesy of Jim Ferreira)

small filament, which appears to sustain an incredible amount of outward pressure until it bursts, spraying material also over a great deal of the Sun. The prominence spray is seen as the less vigorous of the two events (Fig. 2.42).

Limb Flare

Observers are advised to watch for bright knots or mound-like shapes that may appear suddenly at the limb. These blobs could be the result of a flare happening as an active region is either passing over the west limb or newly

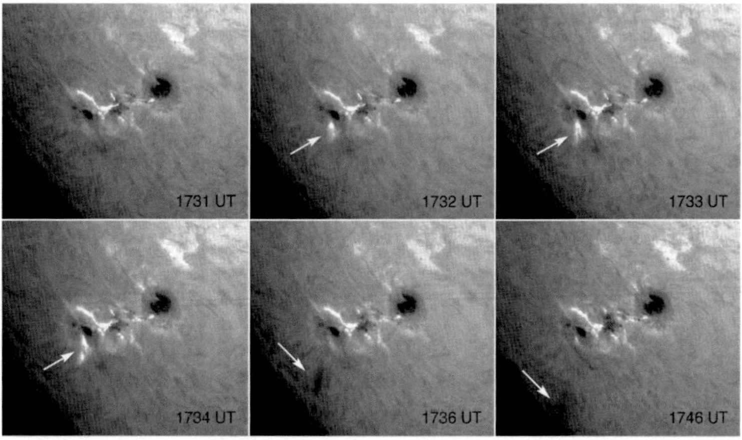

Fig. 2.42 Spray on the solar disc. On 24 September 2011 within AR1302, Jim Ferreira imaged a flare and spray prominence. In the first image at 1731 UT the prominence is not evident, however 1 min later at 1732 UT the eruption is visible as an emission event (material is bright), by 1736 UT the ejected material has become less energetic, and appears as a dark absorption feature. It is speculated that the particularly powerful ribbon flare (*upper right* to spray) is the precipitator to the event (Courtesy of Jim Ferreira)

coming around the east limb. Either situation is ideal for observing post flare loop development or viewing the ejection of material as a puff, surge or spray. Differentiating between a limb prom and a limb flare is accomplished by observing any movement of the feature. A prominence will change position and appearance while a flare will likely just change intensity (Fig. 2.43).

Coronal Mass Ejection

The *coronal mass ejection*, abbreviated CME, is a huge bubble of coronal gas flowing out of the Sun, and propagating as a shock wave in the solar wind. Reaching the Earth in several days, a CME is responsible for geomagnetic storms and their consequences, such as aurora displays and communication disturbances. The content of a CME bubble is plasma (a soupy mix

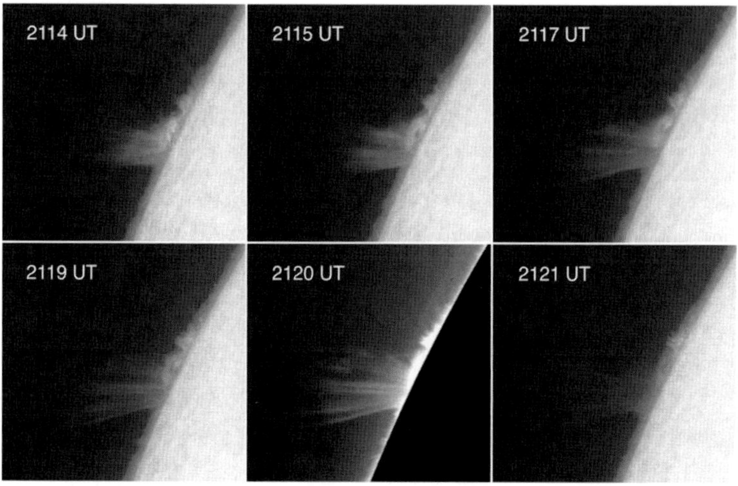

Fig. 2.43 Limb flare and a 48,000 km high surge prominence appear on the Sun's east limb on 17 August 2012. The image from 2120 UT is masked to remove the overexposed disc permitting greater detail of the prom to be visible (Courtesy of Jim Ferreira)

of electrons and protons), but sometimes it may incorporate lesser quantities of other elements, including helium or iron. Events follow the 11-year solar cycle in frequency with an average of more than three bursts a day at solar maximum, and at solar minimum barely more than one per week.

There are three constant attributes of an ordinary CME: a luminous leading edge, followed by a less dense cavity region that embodies the third attribute, a radiant core. It is believed that the core is composed of the material from an erupting prominence (see disparition brusque).

Discounting those unknowingly glimpsed at some total eclipses, the earliest optically perceivable CME occurred on 14 December 1971. Scientists using the OSO-7 (Orbiting Solar Observatory) of the Naval Research Laboratory employed a spaceborne coronagraph to witness the event. Making use of an external occulting disc, a coronagraph blocks light from the Sun creating an artificial eclipse. It is the "eclipse" which allows viewing the feeble light of the Sun's corona.

The eruption at the southeast solar limb from the 1971 event obtained a velocity in excess of 1,000 km/s and was labeled at the time, a coronal transient. *Coronal transients* are any short-lived changes in the structure of the corona. Characteristic expulsions average 400 km/s, and carry off nearly 10 billion tons of solar matter. Further results from the 1970s OSO-7 and Skylab missions confirmed that all prominences that attain an altitude of nearly one-third the solar radius culminate with a CME.

Often a spectacular coronal mass ejection erupts at the solar limb. However, an interesting effect, given the name *halo*, is produced if the CME is directed toward Earth. Through the coronagraph the CME is seen to expand in a radial manner about the solar disc forming a halo or donut shape. When this happens a magnetic storm is a certainty within a day or two. The first visual confirmation of a halo was on 27 November 1979.

Active regions are modeled as one source for the majority of CME phenomenon; the Sun's quiet regions while still a place of origin are less so. Magnetic reconnection is ascribed to both the CME, and solar flare activity. The magnetic fields located in some active regions are intense enough to trap plasma within the arcades (see post flare loops). When reconnection at a lower level within the arcades takes place, a sudden release of energy comes about. The helix of magnetic field left behind from the older connection and the trapped plasma rapidly expands generating an expulsion that possibly hastens a CME.

Representative observations by amateur solar observers are similar to the illustration presented by Steve Rismiller during a ½ h session on 5 March 2000 of an erupting prominence and resultant CME. By 1624 UT the leading edge, cavity, and radiant core are observable well off the solar limb (Fig. 2.44).

Menzel-Evans-Jones Classification System

There is no system of classification that can fully describe all the possible forms a prominence can assume. Regardless, a number of attempts have been made focusing on various attributes of prominences from spectral characteristics to whether they are stable or unstable. The first such system

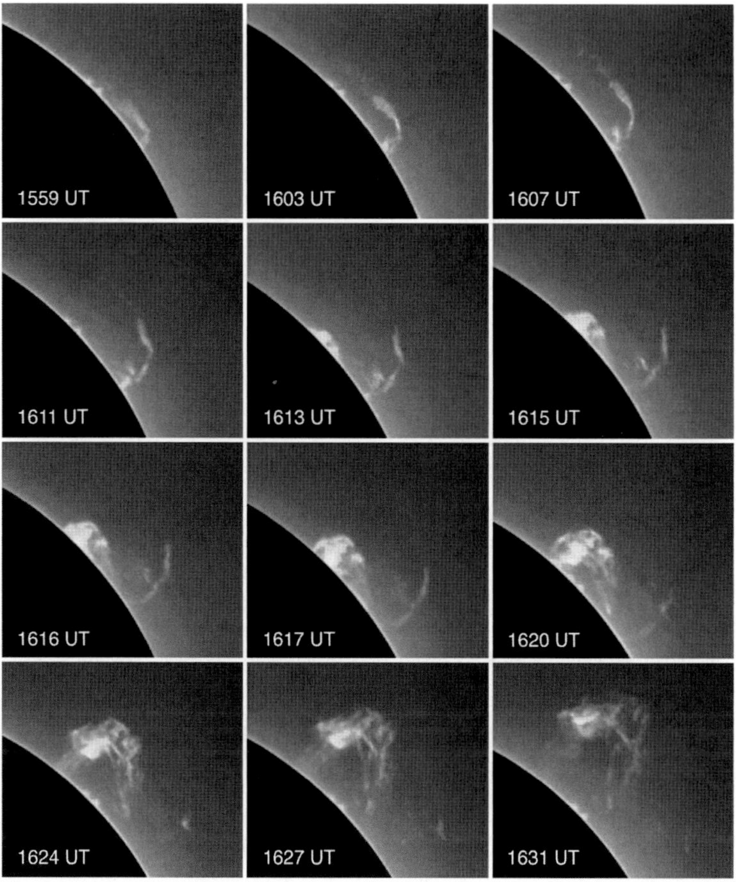

Fig. 2.44 An erupting prominence and coronal mass ejection were caught on 5 March 2000 by veteran solar observer Steve Rismiller (Courtesy of Steve Rismiller)

was devised by Fr. Secchi in the latter 1800s dividing proms into quiescent and eruptive types.

D. Menzel, J. Evans, and F. Jones in the mid-twentieth century introduced a system that separates prominences according to whether they originate in the chromosphere or the corona, and if they are attendant to sunspots.

A simple yet efficient way of cataloging commonplace prominences, the amateur astronomer wishing to maintain metrics on prominence activity is encouraged to consider this plan.

The methodology is similar to, yet pre-dates the McIntosh sunspot classification format in the use of a three-letter designation. In the Menzel/Evans/Jones grading system, the first letter represents the place of origin of the prominence, that is whether descending from the corona (A) or ascending from the chromosphere (B). Since descending or ascending may not be immediately evident, an extended observing period will be required to determine in which direction matter is flowing. Look for knots or blobs, movement of streamers or threads. Motion in a few instances will be obvious, such as with the surge or coronal rain; infrequently horizontal movements do occur, an example being between two quiescent "tree" formations. This movement of matter from one prom to another is called *interaction*, and in the Menzel-Evans-Jones system both proms would be classified as a single hedgerow.

The majority, about 90 % of prominences will be of type A with a downward flow of material. These will have their shape maintained by local magnetic fields. Class B prominences originate in the chromosphere and rise into the corona, often unexpectedly. A prominence can change classes should it become activated.

The second letter of the classification system tells us if the prominence is related to a sunspot or not. Classes AS and BS is an active prom, whereas the AN and BN are quiescent. It may be necessary to make repeat observations for the purpose of ascertaining whether a prom is associated with a sunspot emerging around the east limb of the Sun. Equally for classification reasons, it is important to be familiar with spots disappearing around the west limb.

The final digit in the Menzel/Evans/Jones system denotes the form assumed by the prominence being classified. Rain (ASa) in this system references that which occurs in conjunction to post flare loops (ASl). Menzel's coronal rain (ANa) is defined as that in which streamers originate high in the corona, and persist for several days. Funnel shaped proms; including tornadoes would be class ASf. Types AN with sub letters b, c, and m are filamentary quiescent variations. Those resembling trees (ANc), tree

Table 2.3 Menzel/Evans/Jones prominence classification system

A-Descending coronal prominence	
S-Spot prominence	N-Nonspot prominence
(a) Rain (flare loop associated)	(a) Coronal rain
(f) Funnel	(b) Tree trunk
(l) Loop	(c) Tree
	(d) Hedgerow
	(f) Suspended cloud
	(m) Mound
B-Ascending chromospheric prominence	
S-Spot prominence	N-Nonspot prominence
(s) Surge	(s) Spicule
(p) Puff	

trunks (ANb), and mounds (ANm), may actually be hedgerows (ANd) positioned at an angle parallel to the line of sight. Suspended or detached clouds (ANf) often appear to be completely disconnected, and floating freeing in the corona. There possibly may be connecting streamers, but they are difficult to discern (see Fig. 2.38a).

The surge (BSs) class is inclusive of the spray; this system was not designed to distinguish between the two. A puff (BSp) is an individual knot of incandescence, which erupts at surge velocities. The puff then fades or falls back to the solar surface along the same path of ejection. Spicules are prominence like features, which at times reach heights near 10,000 km. In the Menzel/Evans/Jones system *peculiar* spicules are classed as BNs (Table 2.3 and Fig. 2.45).

Solar Flares

A flare is the extreme, sudden release of stored up energy found most often in the magnetic field of an active region. The violent release of energy happens as reconnection occurs between established magnetic field lines and emerging regions of lower energy flux. Reconnection, also known as realignment, generates massive electrical flows that in turn produce severe thermals. As the surrounding plasma is heated to millions of

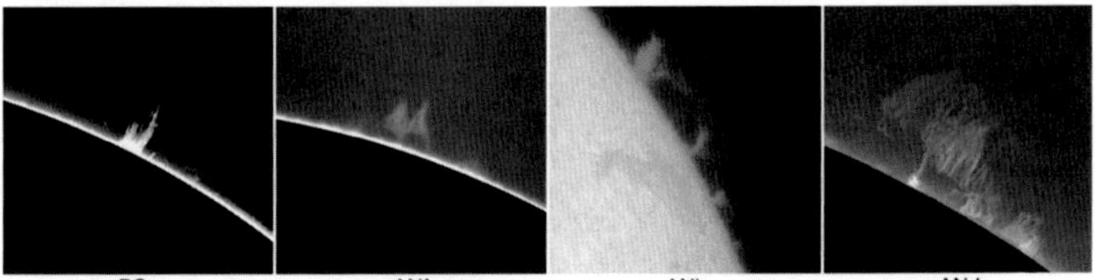

Fig. 2.45 Examples of the Menzel/Evans/Jones prominence classification system. From *left* to *right* is a spray (surge) classed as BSs, second is a suspended cloud, ANf, third is several tree trunks, ANb, and fourth is an activated hedgerow, ANd (Spray image courtesy of Jamey Jenkins, ANf and ANb courtesy of Alexandra Hart, and hedgerow image courtesy of Fabio Acquarone)

Kelvin, the charged particles (electrons, protons, and other heavy ions) as well as the residual magnetic field left over from the preceding connections are powerfully accelerated away from the Sun. A flare has the firepower of packing into a small area on the solar disc (0.01 % of the total) an equivalent of many millions of 100-megaton nuclear bombs. Often times a large-scale flare triggers a coronal mass ejection.

The bulk of flare energy is released at wavelengths beyond those in the visible spectrum. Hard X-rays, EUV (extreme ultraviolet) and microwaves are predominant. A flare visible in H-alpha will show itself as an unexpected brightening of about twice the intensity of the surrounding chromosphere.

Solar flares are not limited to H-alpha viewing. Flares can be seen in the light of calcium and although rare, when spectacularly intense in the solar continuum. Richard Carrington and Richard Hodgson made the first sighting of a flare in 1859, on a projection screen intended for sunspot observing. With the invention of the spectroheliograph some years later, monochromatic observing was made possible, and the nature of flare activity was at last revealed (Fig. 2.46).

Flare Morphology

Visually the typical flare begins as a rapid brightening of several points or *kernels* having a diameter less than 1 arc sec. These are located in most

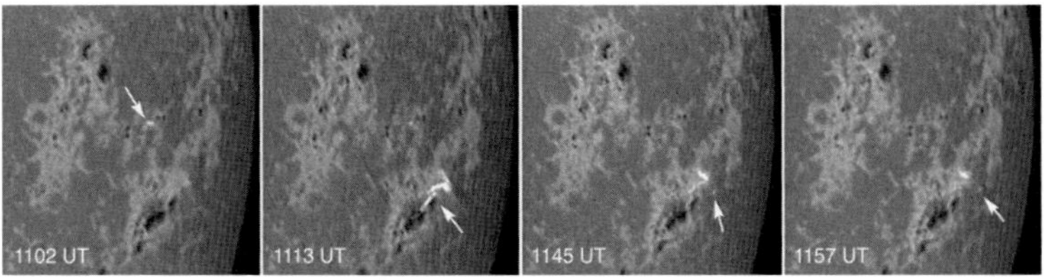

Fig. 2.46 Flare in Ca-K. Here is an illustration from 15 September 2001 of the visibility of a flare in the light of calcium. In the first image note the small subflare (*arrowed*) that is fading. Eleven minutes later in the second frame the larger compact (possibly a single ribbon flare) has attained maximum intensity. By 1145 UT the compact flare has ejected a faint filament (*arrowed*), and at 1157 UT the filament has moved farther toward the Sun's limb as the larger flare continues to fade (Courtesy of Christian Viladrich)

instances inside or near a sunspot. The kernels will grow in size and brightness; for a large flare it may take only several minutes, smaller flares possibly even less. This is then succeeded by a period of rapid brightening and growth, culminating with a peak output followed by a gradual slow retreat in intensity and then decay. There are four recognized stages of flare evolution:

- **Pre-flare/Precursor Phase:** Visible in faint X-rays, where a slow heating of the coronal plasma in the flare region occurs. This phase comprises the activation process for the release of magnetic energy. During a large event the Pre-flare phase lasts roughly 10 min. In H-alpha it makes its presence known by unusual movements of the neutral line filament, and the appearance of new emerging flux.

- **Impulsive Phase:** The rapid release of energy and particle acceleration takes place; notably there will be an increase in hard X-rays, EUV and microwave radiation. The impulsive phase of a large event lasts about a minute. Sometimes there is no impulsive phase or relatively little particle acceleration, this situation gives rise to what is known as a *thermal flare*, and exhibits a gradual rise to peak output followed by an equally slow decline.

- **Flash Phase:** In H-alpha light, the period of rapid brightening to maximum intensity, often several times the brightness of the nearby

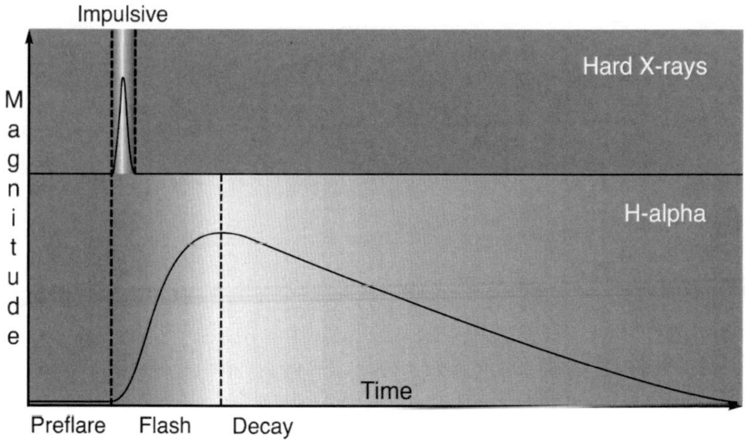

Fig. 2.47 Typical phases of flare intensity in hard X-rays and H-alpha light with magnitude (intensity) vertically graphed against horizontal time. The hard X-rays are expended during the impulsive phase, which occurs during the early moments of the flash phase

chromosphere. Depending on the type of flare event (i.e. subflare, thermal, etc.), this phase may last from several minutes to several hours.

- **Main/Decay Phase:** The gradual decline of H-alpha intensity and a return to the normal state (Fig. 2.47).

For the perceptive Sun-watcher there are specific signs to note that a solar flare could soon be materializing. One preliminary indicator is the classification of a sunspot group. For instance, a mature group of either McIntosh classification D, E, or F is particularly notorious for flare production. Also, keep an attentive eye toward a sunspot for unusual motion, or the emergence of new magnetic flux in a group. The unexpected motions indicate a twisting and distorting of the neutral line within a complex sunspot group, often producing flare output. Other well-known certainties regarding solar flare prediction include:

- The majority of large flares appear 2 years after sunspot maximum.

- Most flares happen as a group develops rather than when a group decays.

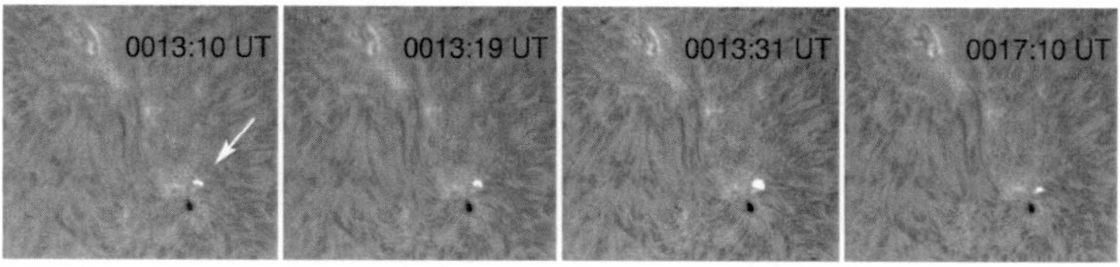

Fig. 2.48 Pictured is a compact subflare in H-alpha from 17 September 2010 near the edge of the penumbra of a small sunspot. The outburst in this case took less than 40 s to reach maximum intensity. Total time transpired to return to original brightness, 240 s (Courtesy of Jim Ferreira)

- Flare production is largely near the same position in a sunspot group.
- Flare activity happens near an active region's neutral line.
- Small flares develop as bright points near a spot on either side of a neutral line.
- Simple groups of bipolar and unipolar classes rarely produce any flares.
- Energetic flares create parallel ribbons on both sides of the neutral line.
- Small flares and a brightening of nearby plage often precede a large flare.

Morphologically, most solar flares fall into two different categories, *compact* and *major*. Compact flares by shear number make up the majority of all solar flare events. They are as their name suggests, closely packed yet intense, some variations being so diminutive they are called *subflares* possessing an area equal to or less than 2.0 heliographic square degrees. Compact flares develop from the brightening of one or more stable loops within an active region. These flares are many times visible in H-alpha at the foot points of an existing loop and occasionally create a surge prominence. Larger compact flares may associate themselves with the complete loop system of an active region. It is not strange for a compact flare to develop near a modest sunspot or a mature unipolar group (Fig. 2.48).

A major flare is a truly large event that always involves the development of two or more ribbons of H-alpha emission. The double ribbon flare, unlike the less intense compact flare, is more climactic, sometimes being linked to the disparition brusque of an overlying ARF.

Complex active regions produce the most violent major flares because of the strong magnetic field present about their sunspots. A major flare can develop near the residue of an old active region harboring a QRF. However, unlike those found in a developing active region, this flare experiences a gradual period of the flash and main phase. This is because quiet solar regions have less powerful magnetic fields than the typical active region.

The development in H-alpha light of a typical double ribbon flare generally follows this order of events:

- At the preflare phase, a region of new flux may emerge near an existing active region filament, which is positioned above the neutral line of an active region. This filament may darken somewhat and demonstrate unusual movements while it begins a leisurely accent. An increase in brightness of the surrounding plage also may be visible.

- As the impulse phase enters, brilliant knots of emission perhaps will be seen near the emerging flux region as well as near the ends of the neutral line filament.

- A rapid progression into the flash phase occurs as the double ribbon develops; one ribbon will grow on each side of the neutral line. The ribbons represent the foot points of the loop arcade that cross the neutral line. The neutral line filament may rapidly erupt as a flare spray, or possibly assume the form of a twisted looping prominence. The peak brightness of the flare is attained.

- During the main phase, the double ribbons move away from the neutral line, normally at 2–20 km/s. As the ribbons separate, the space between them is filled with rising post flare loops that connect the opposite sides of polarity of the neutral line. The loops on the disc may appear dark or bright; on the limb they appear bright. Decay continues as the flare decreases in intensity. When the flare subsides, field transition arches develop across the neutral line, an indication the magnetic fields have returned to a normal state.

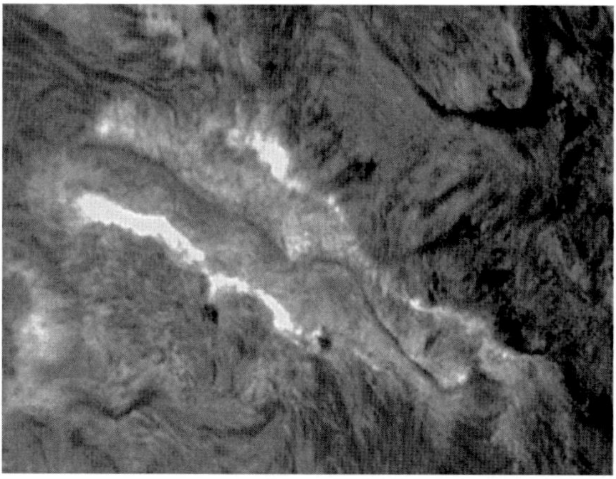

Fig. 2.49 Pictured is a classic example of a double ribbon flare near AR 9085, from 22 July 2000. The view is in the centerline of H-alpha with a 0.56-Å filter. Features to note include the thin winding filament above the neutral line, the pale channel under the filament, and the brilliant ribbons developing on either side of the neutral line. As the main phase develops, these ribbons will separate, one toward the *lower left*, the other toward the *upper right*. Post flare loops will refine across the region occupied by the blown away filament (Courtesy of Gordon Garcia)

Not all major flares necessarily follow this sequence of events to the letter; there may be instances where coronal rain is observed, or the neutral line filament could be absent, or not erupt at all. As with prominences there is a level of uncertainty with solar flares in what may be seen at the telescope at each observing session, hence the allure of observing the Sun (Fig. 2.49).

Hyder Flare

A flare that occurs some distance from an active region, in the quiet Sun, and has a relationship to the sudden disappearance (disparition brusque) of a filament is known as a *Hyder flare*. Charles Hyder, the namesake, was not the discoverer of this type of flare, but rather was the first to make an attempt in explaining how and why it develops.

Visually, the flare shape can assume a series of twisted blobs on one or both sides of a quiet region filament channel, or possibly appear as a one or two ribbon flare. Unlike the active region flare, which is far more impulsive, this event takes an extended time period to evolve. There is little if any particle emission during a Hyder event. The flash phase lasts much longer than an active region flare, up to perhaps an hour, while the main/decay phase can extend to a number of hours. Intensity is less than an active region flare, and is normally spread over a larger area. The overlying filament may simply fade away, or gradually break up and disappear as it rises higher and higher.

Reconnection is believed to be the precipitating cause for a Hyder flare. However, while a number of models exist to explain flare production, the reality is that solar astronomers still do not fully understand how and why they occur.

Moreton Wave

The solar flare causes a number of external events. A brief listing would include: post flare loops, surge and spray prominences, coronal mass ejections, aurora, communication disruptions, and so on. One unique, but very difficult to discern flare related phenomenon is the *Moreton wave*.

A Moreton wave is defined as a shock wave visible in the chromosphere radiating from a large (intense) solar flare. Solar astronomer Gail Moreton discovered these waves in 1959 while working at the Lockheed Solar Observatory. The shock is initiated during the impulsive phase of a major flare and appears as a single arc-shaped form emerging from the active region. A wave is bright in the H-alpha core and dark in the wings. The typical Moreton wave is known to transverse the solar disc at nearly 500–1,500 km/s while traveling 50,000–100,000 km from its source. Disruption of the wave usually occurs if a magnetic barrier is encountered, such as another active region.

Moreton waves are difficult to observe visually because the feature is diffuse and of low contrast. Time-lapse video clips seem to be the best means for identification. By speeding up the progression of the event and its effect on surroundings, the moving arc of a Moreton wave becomes apparent. Occasionally a wave may encounter a filament which in turn reacts like a person on Earth encountering an incoming ocean wave, that is to first be lifted up, and then to be deposited back on the ocean floor. The filament in this case will appear to blink in and out of visibility, the consequence of the rapid up and down motion caused by the passing wave. This development is known as a *winking filament*. Motion of the type described may cause an activation and later eruption of the filament.

Classification of Solar Flares

Solar flares are classified by either H-alpha observations or according to their peak flux in X-rays. For the solar observer intent on classifying any observed flare, the H-alpha method is the only means at his disposal, because it is based on criteria the amateur can measure.

The optical classification of a solar flare is derived from the heliographic area of the flare at maximum brightness in H-alpha. Also known as the "Importance Classification", it is a two-digit system, the first number representing the estimated area of the flare in millionths of the hemisphere, or in square degrees. A millionth of a solar hemisphere is 0.02 square degrees. Area is assigned a rating of 1–4, four representing the larger. Some variations of the scale include the letter "S" to designate subflares.

Square degrees on the Sun are found by measuring the extent of a feature's heliographic latitude and longitude, then multiplying the two figures to obtain the area. Measurements are normally estimations; therefore a calibrated filar micrometer or graduated reticule is ideal for visual determinations of length and width.

The second digit is a measure of the flare's brightness. The brightness of the flare is assigned f, n, or b (faint, normal, or bright). As an example, a solar flare may be classified in this system as: "Importance 2n", meaning it has an area between 250 and 600 millionths of the solar hemisphere and is of normal brightness.

Table 2.4 H-alpha classifications of solar flares

Importance	Millionths of hemisphere	Square degrees of hemisphere
S	<100	≤2.0
1	100–250	2.1–5.1
2	250–600	5.2–12.4
3	600–1,200	12.5–24.7
4	>1,200	≥24.8
Subcode for brightness	–	–
f	Faint	–
n	Normal	–
b	Bright	–

Flare classification using this system can be a bit subjective, particularly as a region nears the solar limb, where foreshortening of the feature makes estimation of area all the more difficult. Even if a flare is near the solar meridian a "gray region" exists among the terms faint, normal, or bright and their interpretation by the observer (Table 2.4).

The second and most often referenced classification scheme is a measure according to the peak flux occurring in soft X-rays (1.0–8.0 Å) performed by the Geostationary Operational Environmental Satellites (GOES). This system provides a practical measure of an event's geophysical importance, unlike the H-alpha system. The GOES system measures the relative X-ray brightness of a flare in comparison to the X-ray brightness of the remainder of the Sun. This classification is divided into a letter system of A, B, C, M, or X, each of which is again divided into nine numerical sub-categories from 1 to 9. Each letter represents a tenfold power increase. For example, an M1 class flare has ten times the power of a C1 class flare and 100 times the power of a B1 class flare.

The A-class flare is miniscule having an X-ray output near background levels. A bit stronger, but still quite weak is the B-class flare. A small flare that still has little to no effect on the Earth is the C-class. Causing radio blackouts and geomagnetic storms of minor consequence is the M-class flare. The M-class is a medium sized event.

The major flares are of classification X, providing the greatest output of emission. An X-class flare can result in a dramatic coronal mass ejection, a disruption in communication, and spectacular displays of aurora.

Table 2.5 X-ray classifications of solar flares

Classification	Peak flux in soft X-rays (1.0–8.0 Å in watts/m^2)
A (1–9)	$<10^{-7}$
B (1–9)	$<10^{-6}$
C (1–9)	$<10^{-5}$
M (1–9)	$<10^{-4}$
X (1–9)	$>10^{-4}$

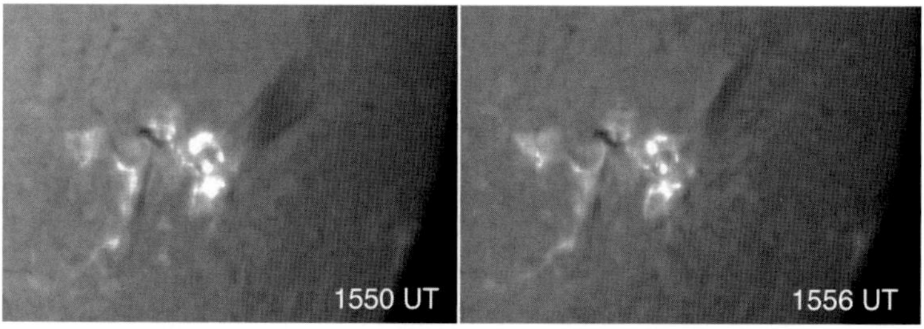

Fig. 2.50 An M3 class flare event from 25 September 2011 in AR1302. Notice the spray prominence (Courtesy of Alexandra Hart)

Whereas the divisions preceding the X class (A, B, C, and M) have nine subclasses of output, the X-class is open-ended. Flares have occurred that exceed X9, including one in 2003 that was beyond X28, the point of saturation for the GOES X-ray sensors (Table 2.5 and Fig. 2.50).

Forecasts for daily solar flare activity and flare classification reports can be found on several Internet websites. One with particularly useful information regarding several areas of geomagnetic activity pertaining to the solar observer is at spaceweather.com. The NOAA, Space Weather Prediction Center at swpc.noaa.gov/index, produces another site with highly detailed information and forecasts.

Corona

Extending further out from the chromosphere and above the transition zone, the gas becomes particularly tenuous. Here is found the pearly white "atmosphere" of the Sun called the *corona*, which means "crown". From the transition zone outward the surrounding temperature begins a conspicuous increase. Within the corona, to a distance of several million kilometers, temperatures exceed 500,000 K and at times reach numbers greater than 2,000,000 K.

The brightest region of the corona compared to the photosphere is nearly one million times fainter, and can best be seen with the aid of a device called a coronagraph, which in principle creates an artificial eclipse of the Sun. Such a device requires special conditions in order to function properly; conditions that reduce scattered light from dust, atmospheric refraction, and water vapor in the atmosphere. While views by the Earth-bound astronomer are limited, superior observations can be made with spacecraft that get above the atmosphere, supporting the coronagraph's revealing qualities.

Since the coronagraph is beyond the capabilities of the typical amateur astronomer, his observation opportunities are restricted to the several minutes of totality that occur during a total eclipse of the Sun. We are fortunate that at times the moon blocks the blinding light of the photosphere, while reducing the light scatter of the atmosphere by the lunar shadow passing from the moon to the ground. If circumstances were different, discovery of the corona would have waited many additional years (Fig. 2.51).

The corona contains various interesting structures, several of which can be seen during totality. The *streamer* is visible in the corona as a bright, thin structure emitted in a radiating fashion all about the Sun. It contains fine rope-like, looping elements reminiscent of the threads in a quiescent prominence. A streamer for the most part makes up the shape of the corona. The *helmet streamer* develops above an active region as a bright looping formation that evolves with a long pointed peak. A magnetic feature, the helmet streamer connects regions of opposing polarity and traps the coronal gas that makes up its form. The more noticeable characteristic, the pointed peak, is created as the solar wind passes between individual streamers. Occasionally a knot of coronal gas breaks away from the tip of a helmet streamer, becoming part of the solar wind. If the gas knot is sufficiently large the result is a coronal mass ejection.

The *polar plume* is also a streamer, but with a much lesser height than other streamers. Located near the north and south poles these figures are

Fig. 2.51 The solar corona visible at the eclipse of 29 March 2006. This image was obtained with a 106 mm refractor through an Astrodon I-series green filter (Courtesy of Christian Viladrich)

associated with a smaller magnetic region on the Sun. Unlike the closed loops of the helmet streamer these structures result from open magnetic field lines. The plume shape is again formed by the activity of the solar wind.

The overall shape of the corona varies with the strength of the sunspot cycle. During sunspot maximum the corona is seen more fully, and is extended about the solar disc, including the high polar latitudes. The streamers seem to point in all directions. At sunspot minimum the corona becomes largely restricted to the sunspot zones, appearing chiefly near the equatorial regions and with fewer streamers (Fig. 2.52).

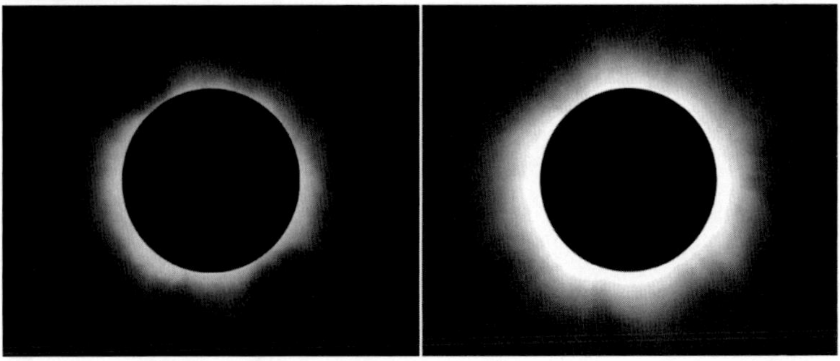

Fig. 2.52 Artist depiction of the corona at solar minimum (*left*) and solar maximum (*right*). During maximum the corona appears uniformly distributed about the equatorial and polar regions, minimum finds the corona less active in the sunspot zones and rather inhibited in the polar regions (Courtesy of Jamey Jenkins)

Solar Eclipse

A solar eclipse, or more technically an occultation, comes in four varieties. The simplest circumstance is the partial eclipse, in which the Moon does not block all of the Sun's photosphere. A partial eclipse happens because the Moon's umbral shadow has either missed the Earth, or the observer is located outside the path of the shadow as it crosses the Earth.

If the alignment of the Sun-Moon-Earth is such that the shadow does intercept Earth, but the Moon's apparent diameter is smaller than the Sun, we have an annular eclipse. An observer located in the path of annularity sees a brilliant ring of light called the annulus, surrounding the disc of the Moon. The annulus is the photosphere and while very interesting, the fringe of light prevents seeing the chromosphere or corona. All phases of any partial or annular eclipse, for safety's sake, must be treated as though one where observing the Sun on any ordinary day. Safe solar filters are absolutely necessary! (Figs. 2.53 and 2.54).

The third variation is the total solar eclipse. Every serious observer of the Sun must make a point of witnessing at least one total solar eclipse. Whereas a partial or annular event may seem blasé, the beauty and awe of

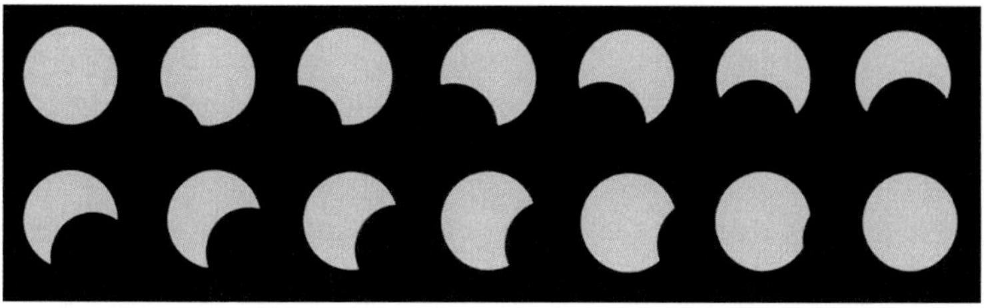

Fig. 2.53 Images of the 25 December 2000 partial eclipse. From the *upper left* to *lower right* all were obtained through a 150 mm f/4 reflecting telescope at 15 min intervals beginning at 1545 UT. In this case the western solar limb is to the *left* (Courtesy of Jamey Jenkins)

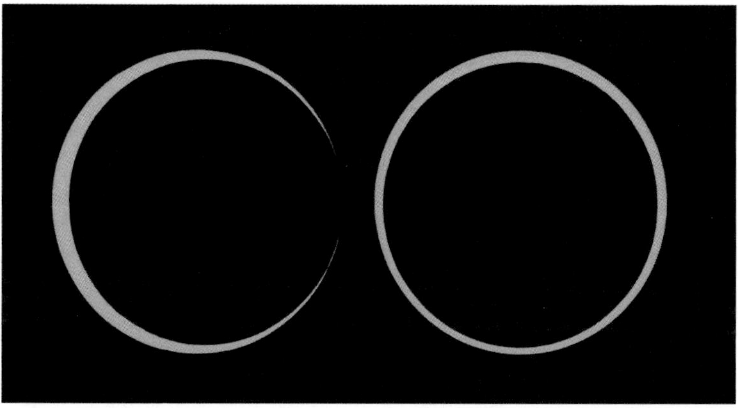

Fig. 2.54 Here is imaged 2nd contact and maximum phase of the annular event of 10 May 1994; solar east on *left*. Kodachrome 25 transparencies attained with a 203 mm aperture f/7 Newtonian telescope (Courtesy of Jamey Jenkins)

totality is inspiring, and a memory never to be forgotten. For the several minutes during totality no special filters or devices are required to see the pinkish chromospheric features or the gleaming silvery corona. A total eclipse of the Sun may afford an amateur astronomer his only firsthand opportunity to observe the corona.

During a total eclipse all the alignments are correct for the Moon to block the photosphere as seen from our vantage spot on Earth. Depending on the circumstances an observer has at most 7 min 31 s to enjoy the spectacle. The typical totality phase lasts on the order of 5 min.

The fourth and most unusual sort of eclipse is the annular/total eclipse, also known as the hybrid, in which the eclipse bounces back and forth between an annular and a total event. The cause of such a mixed up episode is the uneven curvature of our planet. High and low regions of topography affect the Moon's coverage of the Sun contrarily.

Preparations for an Eclipse

Eclipses of the Sun on occasion do happen in the observer's backyard. However, with most events you will find it necessary to travel some distance. The total eclipse pursuer will select an observing site that is located somewhere inside the approximate 100 km wide path of the lunar umbra. Your location could be directly in the mid-region of the path, a position known as the centerline, where the Moon and Sun align flawlessly. Another option to keep in mind is a site within but nearer the edge of the umbral shadow. Such a location often provides long lived spectacular vistas prior to totality, for instance sunlight sparkling through mountain peaks on the lunar limb or extended views of the so-called, shadow bands.

When you travel, sturdy lightweight equipment is called for. High quality imaging can be done with a refractor of less than 100 mm aperture. Purely visual studies can be accomplished with even somewhat smaller telescopes, or lightweight binoculars. Whatever equipment you decide to take, remember that the partial phases can only be seen through safe solar filters. Consider procuring a pair of the special "eclipse glasses" that allow naked-eye observing of the partial phases.

It is vitally important that a study is conducted of the location chosen for viewing the event. This is so as to avoiding unpleasant surprises. What are the chances at that location for cloud cover or precipitation on eclipse day? Is there a convenient spot from which the eclipse can be observed? Will you need and if so have access to a clear horizon? Remember to provide yourself a timepiece that can be set to Universal Time; knowing the accurate time is important for eclipse activities.

Contemplate joining forces with a group of other eclipse chasers. There are several advantages of doing so, for instance a group can guide you on the best practices of eclipse observing, or guarantee the pinpointing of an ideal observing site. Having an experienced eclipse expert at your beckoning may prevent an unexpected mishap to befall a possible once in a lifetime event. Lastly, it's just more enjoyable to observe with a group of like-minded amateur astronomers.

Clothing should be considered for the weather at hand, but follow the same guidelines as when observing the Sun at home. Sunscreen is recommended, as is a long-sleeved shirt and brimmed hat. If you plan on watching the entire eclipse, pack a cooler with drink, sandwiches, or snacks. You may require a folding chair to take a break in, or bring a blanket to spread on the ground for reclining; the partial phases of an eclipse can last quite a long time.

Observing a Total Eclipse

In addition to the above mentioned arrangements, the well-thought-out eclipse chaser will prepare an activity time-line prior to experiencing the event. Because an eclipse is exciting it's easy to become preoccupied, accidentally letting the time slip away while neglecting some observational facet. By preparing a plan of action for the eclipse, nothing is overlooked.

To formulate such a "spread sheet" it's essential to know the predicted times of the various phases called *contacts* of the eclipse for the specific observing location. This information can be gleaned from a small number of online data sources that specialize in eclipse predictions. One excellent site for predictions is provided by NASA at: eclipse.gsfc.nasa.gov/eclipse.html; each prediction includes a link to various tables, maps of the path of totality, and weather prospects at many possible locations. Several additional sites are available if a web search is performed for solar eclipse forecasts.

A sky map showing the location of prominent celestial bodies surrounding the Sun will be helpful for locating Venus, Mercury and any prominent stars that can be seen during totality. Rather than taking written notes, a voice recording device will free up your hands and eyes allowing constant eclipse watching.

Following is a step-by-step discourse of the events that happen at a typical total eclipse of the Sun. Use these insights to plan an observing program that delivers the most from the few minutes allowed to see the coronal solar features.

- First contact begins when the Moon is tangent with the Sun's west limb. This is the beginning of the solar eclipse. The succeeding partial phases will last over an hour, and can optically (naked eye or telescope) be observed **only** with proper safe filtration. The partial phases of an eclipse can also be viewed safely by indirect solar projection using a telescope or pinhole device.

- As the partial phases advance, the local air temperature will begin to drop. Secure a digital or mercury thermometer and take readings at regular 5–10 min intervals, prepare a graph later that relates the before, during, and after temperature readings. Note the habits of birds and animals as they experience the decreasing daylight and temperature; the wildlife will assume night is approaching. Are there many thin crescents on the ground beneath leaf-bearing trees? A cluster of leaves sometimes acts as a tiny pinhole camera each opening projecting an image of the Sun.

- Several minutes before totality watch for the fleetingly seen *shadow bands*. These are narrow shadowy streaks that materialize and travel in various directions. See below for more about observing shadow bands.

- From the west you may be able to spot the Moon's shadow as it glides over the terrain in excess of 2,500 km/h.

- Moments before totality two beautiful effects will be visible at the Sun's east limb. Leading are *Baily's beads*, sunlight passing between several mountainous peaks seen in profile on the lunar edge. The beads will appear as a series of bright points. Then, as the Moon obscures a bit more of the Sun the last bright point is discernible; it is called the *diamond ring*.

- When the diamond ring has disappeared, second contact or totality will have commenced; it is now safe to view the eclipse with your unprotected eyes. Can you see the reddish pink colored chromosphere extending beyond the lunar limb? Are there prominences to be spotted protruding outside the chromosphere? The shimmering corona has details also; is it possible to pick out the variously shaped streamers? Can polar plumes be discerned? What is the general shape of the corona, full and blooming or are certain areas suppressed? Take a brief moment to look away from the Sun, and notice the landscape. Is it as dark as you experience at night during a full Moon? Have the birds become completely quiet? What has the surrounding air temperature reached? Is there a rustling in the trees indicating an increase in wind speed? Briefly study with your unaided eyes or binoculars the sky surrounding the Sun; Venus will likely be the brightest star-like object, can you find Mercury or any brighter stars in the field? Some observers look for comets passing close by the Sun, are any visible?

- The total phase of the eclipse will pass quickly, if photographs are being taken work swiftly and vary the exposures times. Short exposures record the prominences and inner corona; longer exposures record the mid and outer regions of the corona. As totality nears its end, low lying prominences appear on the Sun's western limb. Take care now because the Sun's blinding light is about to burst forth beyond the lunar edge.

- Third contact happens as the diamond ring appears somewhere opposite its previous position, and then Baily's beads which are followed very shortly by the thinnest sliver of the photosphere, precipitating once more the transient shadow bands. **Safe solar filtration is necessary at third contact.**

- The Moon's shadow will exit toward the east with totality over. Leaving a recreation of the earlier partial eclipse until the Moon completely exits the Sun at fourth contact (Fig. 2.55).

Shadow Bands

As totality nears, the Sun's remaining sliver of photospheric crescent becomes progressively thinner *sometimes* creating a unique phenomenon

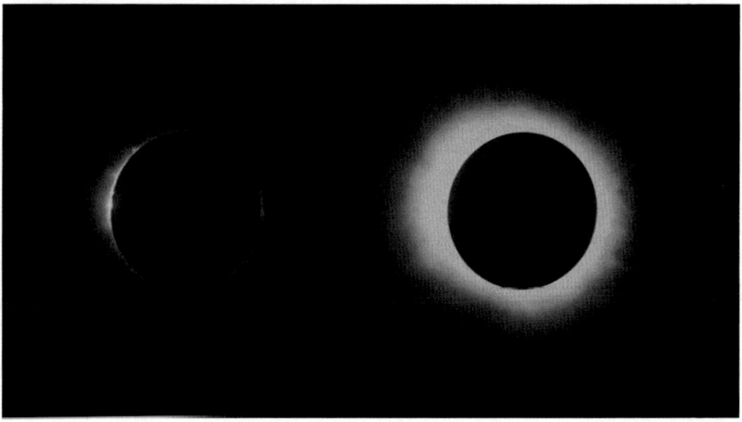

Fig. 2.55 Totality from 26 February 1979 near Brandon, Manitoba, Canada. *On the left* is the last remaining glints of sunlight before totality, a "mini-diamond ring", also notice the *reddish* prominences on the Sun's east limb. In the *right hand image* take note within the corona (3 o'clock position) of an embedded "S" shaped feature termed a sigmoid pattern (Kodachromes courtesy of by Jamey Jenkins)

that we call shadow bands. The spectacle happens just within the last few minutes prior to and following the total eclipse, each occurrence an unpredictable length of time.

Shadow bands are very low contrast, light and dark wavy forms visible on the ground or a lightly colored surface, such as a sidewalk or building. Possessing irregularly narrow widths of 20–50 mm with separations of 50–250 mm, shadow bands are fleeting, sometimes stationary, and at other times having a movement approaching 3.0 m/s.

This interesting effect develops because the thin solar crescent is illuminating lower atmospheric turbulence. Shadow band motion is the result of air mass movement. Somewhat of a novelty, not all experienced eclipse chasers have observed them; shadow bands can therefore serve as a badge of honor when spotted. Photographing a series of moving bands is particularly a challenge.

Potential shadow band observers can create an observing station designed to increase their odds for spotting the feature. All that is really required is

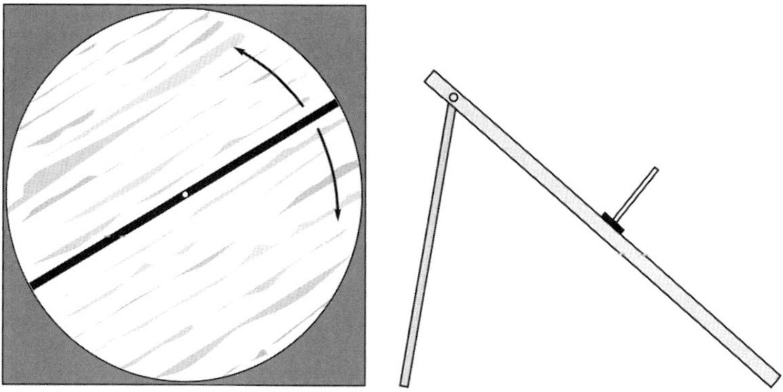

Fig. 2.56 Straight-on and side view drawings of the shadow band viewing screen. The *dark colored bar* pivots about the center, aligning with the illusive bands, and providing a reference for their detection

a flat surface with a lightly colored background, but this described apparatus includes an orientation shaft, and rotating bar for marking the bands' capricious location.

Seen at past eclipse expeditions, the station consists of a stiff white screen formed from smooth plywood, cardboard, or even cloth stretched over a wooden frame. The screen should be as large as is practical, but no smaller than 1.5–2.0 m^2. At the center of the screen a round wooden positioning dowel is fixed, protruding about 300 mm and perpendicular to the panel. Scribed with the dowel at its epicenter is a circle as large as can be fitted on the screen. Adjustable legs to prop the screen at an angle equal to the Sun's altitude during totality are required. If the Sun is at the zenith, the screen would lay flat and no support is necessary.

A straight-edged, dark colored bar with a pivoting hole of the same diameter as the dowel is situated over the positioning rod; the bar is able to rotate freely yet with enough friction so that gravity will not disturb it. The bar will have a length equivalent to the diameter of the scribed circle, and a width of about 25 mm. That describes the basic observing station (Fig. 2.56).

Shortly before totality direct the screen toward the Sun, so that the dowel serving as a pointer casts no shadow. The viewing screen will now be perpendicular to the Moon's oncoming shadow. With several minutes until second contact face the screen, but do not allow your own shadow to fall on it. Watch intently for any sign of the shadow bands. Should they appear, quickly move the bar so that it now rests parallel to the bands and perpendicular to their motion; the straightedge thus aiding in recognition. For directional studies pre-position the bar so that it points east–west, thus indicating the terrestrial direction of shadow band motions. The bands may be dull, nebulous and of low contrast; perhaps appearing stationary or rapidly in motion. Results at any eclipse are always unpredictable.

Photos of the screen with shadow bands are best obtained at a high shutter speed with the images worked later using software to increase contrast. A video illustrating the movements are edifying, and particularly become useful for reducing movement velocity. Following the third contact phase observations should be repeated to see if the shadow bands reappear.

PART II

OBSERVATIONAL TECHNIQUES

CHAPTER 3

OBSERVING THE SUN

Without a doubt you will find the Sun one of the most enjoyable and thrilling observing experiences in your astronomical pursuits. The excitement of turning a telescope skyward and "discovering" a new sunspot or witnessing first hand the eruption of a prominence, and then watching it rain back down on the Sun's surface can be awe-inspiring. The truth is that few other bodies in the heavens offer the sheer dynamics viewed on the Sun.

Since the numbers of solar observers have increased in recent years, largely because of the availability of affordable safe solar telescopes, and high quality accessories that turn conventional scopes into safe instruments for exploring the Sun, it is common to find yourself among like-minded amateurs at any astro-event. The topic of discussion at these gatherings usually centers on what has been dubbed the three "S's" of Sun-watching – which when listed in their level of importance to the solar observer are Safety, Seeing, and Suitable instrumentation.

Solar Safety

The first and most important lesson to be taught with any form of observing the Sun is the inherent danger involved in this activity. If necessary create a placard emblazoned with these words and attach it to any instrument used for solar observing: **NEVER attempt to view the Sun through any optical instrument that has not been properly fitted with SAFE solar observing accessories!** Another lesson that must be understood by all amateur astronomers, whether interested in observing the Sun or not is this: **NEVER stare at the Sun with your unaided eyes, unless looking through a known and tested solar filter intended for such use.** Remember, it only takes a moment of exposure to unfiltered sunlight to permanently destroy your eyesight. There are no second chances given in this activity!

Jamey L. Jenkins, *Observing the Sun: A Pocket Field Guide*,
Astronomer's Pocket Field Guide, DOI 10.1007/978-1-4614-8015-0_3,

Specifically, what are the dangers to avoid when observing the Sun? One is overexposure to light, primarily in the blue-green region of the spectrum. This causes damage to the light sensitive cells of the eye called, the rods and cones. Chemical changes occur when we experience overexposure to bright sunlight, changes that leave us with blindness in our eyes either for a short period of time or permanently. This chemical retinal damage can result from either a single exposure event or the cumulative effects of a number of seemingly innocent "short looks" at the Sun.

Although overexposure to the bright visible light of the Sun is a serious factor in blinding the observer, there are also other damaging rays. The Sun emits invisible radiation in the form of ultraviolet (UV) and infrared (IR) light. Ultraviolet light (280–380 nm), although not reaching the retina due to absorption by the eye, contributes to the development of future cataracts, as well as speeding up the aging process of the eye's outer layers. When infrared light (780–1,400 nm) is admitted to the eyes, heat burns the exposed internal tissues, resulting in destroyed rods and cones. Much of the thermal energy from the Sun is in the form of IR light. A permanently blind region on the retina will be the end effect of this exposure. Unfortunately, the observer may not immediately be aware of the injury, as there is no means of sensing heat in our eyes. The effects are sometimes not evident until after the damage has been done, usually within a few hours.

Never utilize the following items for observing the Sun: CD's or DVD's, mylar food packages, photographic color film, conventional sunglasses, crossed polarization or neutral density filters, and welder's glass with a density less than shade #14. These products used alone transmit unsafe levels of infrared (IR) and ultraviolet (UV) light, which will damage our eyes irreversibly!

Another dangerous product to avoid that occasionally shows up on today's market is the eyepiece solar filter. This is often distributed with a low-end department store telescope and is intended to screw into the barrel of an eyepiece to facilitate direct viewing of the Sun. The filter is no more than a very dense-colored glass. After placement at the focal plane of a telescope directed toward the Sun, the filter *will* eventually shatter from exposure to the concentrated light and heat, exposing the eye to the harmful sunlight. If you acquire one of these filters, discard it immediately!

So, what constitutes a safe white light solar filter? To be a visually safe device, the appliance (accessory that attaches to a telescope) must block IR and UV rays and reduce the intensity of the intervening wavelengths entering a telescope to a percentage of 0.003 % (filter density 4.5). This is the lower limit of safety; many individuals prefer a filter transmitting even less, a filter with a density of about 5.0 is found comfortable by most observers. At this level, the Sun appears no brighter than the full moon.

Observers of the Sun in monochromatic light (usually H-alpha or Calcium-K) find few issues with safety since their equipment, if properly used, excludes all off-band wavelengths of light other than in the specific spectral regions being observed. A specially made pre-filter called an energy rejection filter (ERF) is used over a telescope's objective to absorb and/or reject UV and IR light. This pre-filter protects both you and the expensive narrowband filter. The filtering package contains additional trimming and blocking filters that remove unwanted wavelengths, with the end result being a safe, monochromatic view. The key point to remember with narrowband equipment is to **always** follow the manufacturer's instructions for use of their product.

Whether you are observing the white light or monochromatic Sun, periodically check and double check the filtering systems on your telescope, making certain the components are properly and securely mounted. I have seen equipment become loose and accidentally set ajar or even knocked off a telescope. Avoiding a perilous accident in those circumstances should never be just a matter of luck.

Locating the Sun in your telescope can be an issue, if not properly done. Avoid sighting along the edge of the telescope tube, a technique akin to staring at the Sun. And for safety's sake either remove the finder telescope, or cap its objective to prevent it from projecting a beam of hot focused sunlight.

There are two recommended ways to find the Sun through a telescope. The first and most often used technique is to locate the telescope's shadow on the ground when pointing your instrument Sunward. Watch the shadow become smaller and smaller until it forms as tiny a profile of the scope as possible. At that time the Sun should be very near the center of a low-power eyepiece. A second and even more accurate method is to utilize a pinhole or bull's-eye projection finder. One of these finders may be made in your home workshop or purchased commercially from a telescope dealer. The design is simple often with no more than a tiny hole at one end and a target opposite. Properly align the pinhole finder to the optical axis of the telescope, and whenever the solar image falls on the target, the Sun is located in the telescope.

Personal safety considerations for the solar observer are important, and should not be neglected. Extended exposure to sunlight (UVA/UVB radiation) can be dangerous to a solar observer's health. That is why sunscreen lotions are important for the body. During extended observing sessions use a product with a suitable sun protection factor (SPF rating). The arms, face, and neck seem to be particularly vulnerable. Many observers wear a hat or cap to shade their face from direct sunlight. The cap can also be worn reversed with the bill protecting the neck when looking into an eyepiece. Additionally, a long sleeved loose-fitting shirt offers protection to the arms.

Seeing Conditions

Veteran solar observer Gordon Garcia once remarked that when it comes to observing the Sun, "seeing is everything". After safety, seeing is the Holy Grail to a solar observer. It is the one facet that an observer may have the least control over, that creates the greatest frustration at the eyepiece, and is often the determining factor between observational success and failure. But if specific guidelines are followed, seeing conditions can many times be improved.

What is astronomical seeing? In the end, seeing describes the state of the atmosphere between a celestial object and the observer's eye. The air quality (i.e. temperature, turbulence, wind speed) affects seeing the most, but those are not the only factors. Instrumental performance also plays a role in determining seeing conditions. Air currents within the telescope itself can disrupt the view just as easily. Closed tube telescope designs such as with the refractor suffer the least; the open tube Newtonian reflector can be the most vulnerable. It doesn't take the novice astronomer long at the eyepiece to discover that patience is a virtue when observing the Sun. You may be treated to a tack sharp view one moment and a boiling bubbling blob the next. A circumstance of good seeing conditions is the difference between observing fine solar detail and viewing vague uninteresting smudges.

Seeing can be improved if the following measures are considered in your solar observing routine. Locate your telescope in a place where the prevailing winds are not blocked by buildings, trees, tall fences, hills, or other obstructions. Open air locations surrounded by grass or water are preferable to asphalt or concrete parking lots, which can absorb heat and radiate it into the path of light entering a telescope. Heat waves are only one cause of poor seeing conditions. Atmospheric refraction is a condition amplified when observing an object closer to the horizon. Recall the prismatic effect bright stars and planets display if viewed near the horizon? That is atmospheric refraction and can be controlled by observing the Sun near the sky's central meridian. A colored filter added to the eyepiece also helps remove unwanted refracted light if observing away from the meridian.

Lastly but importantly, study the local weather patterns to discover what conditions prevail during times of best seeing. Things to consider are: approaching cold or warm fronts, the direction and strength of winds, time of day, and altitude of the Sun above the horizon. Learn the circumstances of excellent observing conditions in your locale and when they repeat, take advantage of the opportunity.

If you maintain a system of metrics when conducting an astronomical observation, a scale for determining current sky conditions is required. Estimating the visibility of a specific solar feature can be one method of determining the quality of daytime astronomical seeing. Granulation (1 arc sec resolution feature) in the photosphere typically requires excellent seeing conditions to be clearly visible, poorer conditions render it proportionately less so. A scale representing from 1 to 5 arc-sec seeing can therefore be assigned to the given visibility of granules. For instance, seeing is considered better than 1 arc-sec when granulation is clearly visible, a range of 1–2 arc sec means that granulation appears mottled, and so on. See Table 3.1 for details.

The atmosphere also contains levels of water vapor, dust, smoke, and other particles that affect the clearness or transparency of the sky. When transparency is reduced, solar features begin to lose clarity. The seeing may not be affected, but contrast and brightness can suffer. At one end of this scale you find deep blue skies permitting crisp, clean viewing and at the other end complete opaqueness, as during overcast skies. To describe transparency, simple terms such as excellent, good, fair, and poor suffice (Fig. 3.1).

Table 3.1 Scales for defining seeing and transparency

Seeing scale	Description
<1 arc sec	Granulation resolved
1–2 arc sec	Granulation appears blotchy
3–5 arc sec	Granulation/pores sometimes visible
>5 arc sec	Granulation/pores are not visible, ill-defined sunspots
–	–
Transparency scale	Description
Excellent	Clearest conditions/deep blue sky
Good	Clear but not a good as perfect
Fair	Hazy sky
Poor	Thin clouds covering Sun

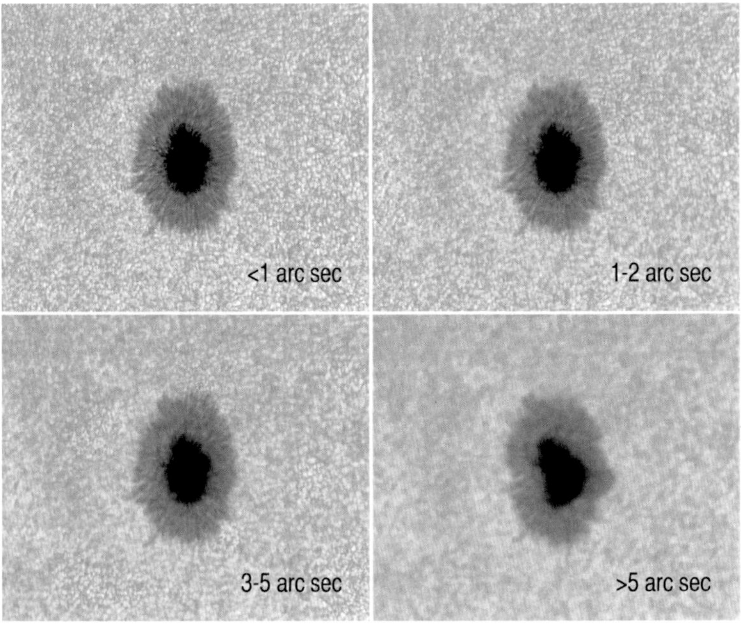

Fig. 3.1 Visual depiction of the arc second seeing scale. Ideal seeing is <1 arc sec and a barely acceptable condition would be >5 arc sec (Courtesy of Art Whipple)

Instruments for Solar Observing

Casual solar observing can be accomplished with a number of instruments ranging from eclipse glasses to a large compound telescope. However, the serious observer will most often select an instrument package that delivers optimum performance for the type of studies he carries out. The best *all-around* telescope for observing the Sun in white light, Calcium-K, or Hydrogen-alpha is the refractor.

It takes more than just a telescope to observe the Sun. A number of appliances are important: a projection screen or objective filter for white light observing and for monochromatic observing, a narrowband filter are a few of the typical add-ons.

Reflecting Telescopes

There exist several telescope designs that utilize reflecting optics; the most common is the traditional Newtonian reflector, invented in the seventeenth century by Isaac Newton. One reason a Newtonian is popular with a telescope maker is because of the fewer optical surfaces required to figure. A 150 mm f/8 configuration is quite common within the amateur ranks.

Advantages: more cost effective per aperture than other telescopes, totally achromatic performance, long focus instruments are of superior quality.

Disadvantages: open-ended tubes generate tube currents (truss-tubes can help to eliminate the problem), awkward when using solar projection or end-loading monochromatic filters, central obstruction limits resolution/contrast.

Compound Telescopes

Cassegrain, Maksutov, and the Schmidt telescope all fall into this category. These telescopes utilize several mirrors or a mirror/lens combination to form the primary image. The compound telescope is a favorite all around astronomy choice because of the portability factor offered in a compact package. Frequently found telescopes are 200 mm aperture with an f/10 focal ratio.

Advantages: long focal length in a compact tube, often a sealed tube with no tube currents, extremely portable, plenty of back focus for accessories.

Disadvantages: not suitable for white light solar projection because of the damage risk to internal parts, fork mounting is not best option for solar projection, large central obstruction, odd light path angles result when used off-axis.

Refracting Telescopes

A refracting telescope has a lens as the primary optical component to gather and focus light. If the objective lens is made of two elements (usually crown and flint glass) it is called an achromat. A more complex design with three or more lens elements is called an apochromat. The apochromat minimizes color errors producing a view without color fringing, similar to mirror optics.

Advantages: no central obstruction for sharp/high contrast views, plenty of back focus for accessories, most convenient for solar projection, most practical design for monochromatic observing, closed tube limiting internal tube currents, portable in medium focal lengths.

Disadvantages: cost is greater per aperture, long focal length instruments can be unwieldy, avoid plastic internal parts which can melt if exposed to heat from the Sun.

Dedicated Solar Telescopes

The astro-consumer is given a number of choices when it comes to an available pre-packaged solar telescope. There exist white light models, those engineered for H-alpha observing, and an equal number of Ca-K telescopes.

By far the prototype is a refractor, sporting a modest aperture and focal length. Front-loading etalons are nearly standard, although some designs have moved the filter package to the interior of the telescope.

Considerations when purchasing one of these telescopes include: what you wish to observe (WL, H-alpha, or Ca-K features), aperture, bandwidth *and* cost, since those four are closely related. If you are interested in only casual, low power wide field views, then a 40 mm aperture telescope may suit your needs. Close-up detailed studies will require something a bit larger, in the 90 mm and greater range. Likewise, a monochromatic view with ample contrast requires a bandwidth filter of less than 1.0 Å. If none of the dedicated solar telescopes are able to meet your needs, investigate adding specific appliances to your existing telescope that permit white light or monochromatic observing.

White Light Appliances

There are two techniques for observing the white light Sun: the direct and indirect methods. Indirect observing is slightly safer and is known as solar projection. This technique was developed hundreds of years ago by the earliest solar observers. It involves using the eyepiece of the telescope as a projection lens, enlarging the Sun's image onto a white screen positioned some distance behind the eyepiece. Projection screens can be purchased or homemade. Important features of a screen are as follows: the screen must securely attach to the projecting telescope; it should permit adjustment in the projection distance, and have a means to control ambient light from spilling onto it. Ambient light has a tendency to wash out the lower contrast details of the Sun, features like granulation and faculae. Additionally, the finish of the projection screen must be smooth; any roughness will hide the finer solar details. While solar projection has limitations as above, some observing projects, like positional studies are best performed with this technique.

Table 3.2 provides a reference for projection distances when using one of several common eyepieces. The standard projected size of the Sun is 150 mm. Begin by calculating the approximate size of the Sun at the focal

Table 3.2 Projection distances with common eyepieces. A screen with adjustable distance is necessary to correct for instrumental variables and the changing Earth-Sun separation found throughout the year

–	Eyepiece focal length				
Projection magnification	12 mm	15 mm	20 mm	25 mm	30 mm
5X	72 mm	90 mm	120 mm	150 mm	180 mm
10X	132 mm	165 mm	220 mm	275 mm	330 mm
15X	192 mm	240 mm	320 mm	400 mm	480 mm
20X	252 mm	315 mm	420 mm	525 mm	630 mm

plane of your telescope using the formula: telescope focal length (mm) × .009 = diameter of Sun (mm). Divide 150 mm by that diameter to find the required projection magnification. Refer to the Table, locate the eyepiece focal length and the computed magnification to determine the required distance between the eyepiece and projection screen. Odd magnifications will require interpolating some distances. The ability to adjust the distance between eyepiece and screen is put to advantage here.

When projecting the Sun a telescope's aperture is normally masked to 100 mm or less, so as to curtail excessive heat at the focal plane. Select an eyepiece constructed without optical cement that could possibly melt and destroy the lenses in the focused heat of the Sun. Typical projection eyepieces include the elementary Ramsden and Huygenian designs. Telescope owners having instruments of less than 76 mm aperture that generate a minimum of heat, occasionally use a higher quality eyepiece (i.e. orthoscopic or plössl) for sharper projection images. Regardless of the telescope/eyepiece combination a good rule-of-thumb is to periodically turn the instrument away from the Sun, which allows the optics to cool a bit before resuming observations.

An alternative to the reflecting screens illustrated in Fig. 3.2 is the adaptation of a translucent projection screen material creating a unique direct view. Again ambient light unless curtailed will reduce contrast, and the texture of the translucent screen must be considered regarding final image quality. Nevertheless interesting results can be achieved, making the technique worth investigating.

Fig. 3.2 Homemade projection screens for casual sunspot observing. The box shaped example on the *left* was spliced together using discarded foam board, epoxy, and spare telescope parts, built by Raymond Rienks. The second illustration is a Hossfield Pyramid, named for AAVSO solar observer Casper Hossfield. Jamey Jenkins constructed this version from black foam board, masking tape, and white card stock

Direct observation through a suitable objective solar filter is today the most widely used mode of white light observing. The objective filter securely attaches over the entrance of the telescope. Be sure the filter is a *visual* model rather than a filter intended for photographic use. A photographic only filter has less density, therefore transmitting more light. At this writing the best bargain, providing excellent quality images is the Baader Planetarium AstroSolar™ Safety Film.

An observer must prepare for the potential of a filter being "knocked off" the telescope, by human error or an unexpected gust of wind. Securing the filter with nylon screws is sufficient, but always double-check that the retaining screws are snug. For an additional margin of safety, wrap the edge of the filter's cell with masking tape thereby temporarily bonding it to the telescope tube.

The least used direct method, but arguably providing the highest quality view, is the adaptation of a Herschel wedge and internal supplementary filters. The Herschel wedge is a thin prism-shaped piece of glass, which reflects about 5 % of incoming light from the Sun to a telescope's eyepiece. Additional filters are inserted between the prism-shaped wedge and the eyepiece to dim and/or accentuate specific wavelengths to a safe visual level. Some observers swear by the Herschel wedge, and while is a safe viewing appliance, I usually express reservations because of the heat allowed to enter the telescope prior to any filtering of the image. Heat can damage internal telescope parts, and is a source of air currents that contribute to poor seeing conditions. The wedge user will counteract this by allowing the telescope to warm to an "operating temperature", consequently reducing tube currents.

A colored glass filter, the type used by an observer of the planets is also helpful for enhancing features of the white light Sun. You must adhere to this common sense rule when using supplementary filters: **No eyepiece filter is ever to be used alone for solar observation; it should only be used to supplement a safe solar filter already in use.**

Glass eyepiece filters are available in a myriad of hues and shades. The most helpful colors for a solar observer will be blue, green, and red. Briefly, let me say that a blue filter enhances faculae that are located some distance from the solar limb. A green filter improves the contrast of faculae and granulation and a red filter enhances sunspot umbrae and also increases the contrast of penumbral streaks.

An ordinary glass filter works by the principle of absorption. Light passing through the filter has some wavelengths absorbed by internal aggregates, while other wavelengths pass right on through. The bandwidth of a filter is quite wide, on the order of several 100 Å. Some manufacturers produce interference type filters similar to the glass variety that also enhance white light features. These filters have a narrow bandwidth, providing superior performance. The bandwidth of such a filter is not narrow enough to show prominences or other monochromatic light features, but a filter near 520–540 nm significantly improves granulation and facular visibility. The Baader Calcium K-Line filter (8 nm bandwidth centered on 395 nm) is useful for noteworthy Ca-K features, and a filter passing light around 430.5 nm (G-band) is serviceable for locating white light flares in the solar continuum.

Table 3.3 Color filter usage for solar observing

Color	Wratten filter	Application
Dark Red	29	Red and orange filters increase the contrast of knots and radial streaks in the penumbrae of sunspots
Red	25A	–
Light Red	23A	
Orange	21	–
Yellow	11	Yellow, an all-around filter that lessens achromatic color errors
Light Green	56	Green filters enhance the visibility of granulation and faculae
Green	58	–
Dark Green	61	–
Blue	47	Blue will assist in seeing faculae further from the limb than normal

The K-Line and G-band filters are most effective if used photographically. Of course, these interference filters are used only to supplement a safe solar filter. See Table 3.3 for a listing of color filter usages.

Monochromatic Light Appliances

The intense brilliance of the photosphere outshines the feeble weaker light of the Sun's chromospheric atmosphere. This is why we cannot see features such as the prominence with our conventional telescope setup. And so in the past, astronomers were limited to a total solar eclipse for brief studies of any chromospheric limb features. With the advent of the spectroscope a means for studying disc and limb phenomena any time of the day was realized. And then since the 1930s, technological improvements have gradually allowed for the development of filtering systems that show the solar chromosphere with a bandwidth as narrow as .1 Å. An amateur solar astronomer ordinarily observes the monochromatic Sun in either the light of hydrogen-alpha (H-alpha or Hα) at 656.3 nm or at 393.3 nm, the K-line of singly ionized calcium (Calcium K-line or Ca-K). The significance of these filters for an astronomer becomes obvious by their convenience, adaptability to existing telescopes, and the superior performance demonstrated as compared to spectrographic equipment from the past.

Today's modern monochromatic filter is available in two basic designs, with a few unique variations to be found in the marketplace. The two approaches are called front-loading and end-loading, a reference to the location of the filter on a telescope. A front-loaded filter uses a large diameter etalon placed outside the entrance to the telescope. An etalon is a filter design that operates by the multiple-beam interference of light, reflected and transmitted by a pair of parallel flat reflecting plates. In a nutshell, it creates the thin slice of monochromatic light necessary to see chromospheric features. An end-loaded system uses a smaller diameter etalon and positions it inside the light path of a telescope, just before the eyepiece.

Coronado Technology Group entered the amateur marketplace in the 1990s with their introduction of sub-angstrom (<1.0 Å) front-loaded H-alpha filter systems. Since that time several manufacturers have introduced their own lines of front-loading filters. These devices combine a necessary ERF (energy rejection filter) that removes unwanted UV and IR light, and the large diameter etalon in a compact package, reminiscent of a white light objective filter. Similar Ca-K filters are also available.

All sub-angstrom filters are extra-sensitive to the angle of light passing through them; non-parallel light beams or non-perpendicular (relative to the filter) beams take a physically longer path through a filter, resulting in an altering of the filter's expected performance. The ability to adjust a filter's tilt (tunable) allows the observer to correct for instrumental and observational miscues. Temperature also affects a sub-angstrom filter's performance, expanding or contracting the spacing of the elements within the filter pack. Consequently, a recommended operating temperature is given for maximum performance of most interference filters. Etalons are also known to transmit "side bands", bits of the spectrum other than the desired output wavelength. To remove the unwanted side bands from a front-loaded etalon, a so-called blocking filter is often inserted before the eyepiece.

All things considered, front-loaded appliances are indeed excellent accessories, convenient to use and very portable. There even exists a front-loaded technique called "double stacking"; whereby a second etalon is mounted in conjunction with the first unit to narrow the total bandwidth, increasing contrast of a feature.

End-loaded filter systems are of two classes: the sub-angstrom model and what is commonly known as the "prominence viewer". *Either* class requires the use of a separate ERF mounted over the entrance to the telescope. Oft times an ERF is of a smaller diameter than the telescope aperture, resulting in an f/20 focal ratio for a prominence viewer, or an f/30 ratio for a sub-angstrom filter package.

Prominence viewers with a bandwidth greater than 1.5 Å and less than 10 Å also require the addition of an occulting cone or disc to block the intense brilliance of the Sun when observing limb features. The solar disc appears brighter through the prom-viewer because the wider bandwidth of the filter allows light from either side of the H-alpha line (called the solar continuum) to pass through to the eye. This is not a device recommended for the novice; effective use requires a telescope drive system with accurate slow-motion controls.

A sub-angstrom end-loading filter requires near-incident (nearly parallel) light beams to function properly. The near-incident beam may be obtained by masking the telescope objective to an effective f/30 aperture (sub-diameter ERF) or by inserting into the optical train a series of lenses that force the light beam from the objective to near incidence. Such a device is called a telecentric. A telecentric ensures that a full aperture ERF may be matched with the telescope, allowing full telescopic resolution.

Normal operating temperature for most end-loading systems is between 90 °F and 150 °F, in which case an electrically powered heating unit (oven) is installed around the etalon. Decreasing or increasing the oven temperature regulates the transmission characteristics of the filter by about an angstrom on either side of the desired output. When an observer intentionally alters the oven temperature (also a means of tuning), causing the filter to shift off-band, this is called viewing in the wings (i.e., blue-wing or red-wing of H-alpha). Observing in the wings allows scrutinizing of a lower region in the chromosphere, or viewing a feature that has been Doppler-shifted and is otherwise invisible to the observer on-band. Unfortunately temperature changes are slow to initiate. A number of observers have constructed mechanisms that allow rapid tilting of a filter relative to the incoming beam of light, a much faster means of viewing in the wings. Side band wavelengths are also experienced with an end-loaded etalon. Again a blocking filter is included in the filter pack to exclude these unwanted wavelengths.

H-alpha observing provides a profusion of observational activity in the chromosphere, but a filter system designed to transmit light of the H (396.9 nm) or K (393.3 nm) lines of singly ionized calcium found in the violet region of the spectrum, shows the cooler region immediately above the photosphere. A filter having a bandwidth of 2.0–4.0 Å is sufficient for Ca-H or Ca-K studies since the H and K lines are spectrally broader than the H-alpha line, and therefore still nearly isolated from continuum light.

Other Observing Accessories

A means of controlling brightness at the eyepiece is a valuable addition, particularly for the H-alpha observer. Insert a pair of adjustable polarizing filters before the eyepiece of the H-alpha scope. Rotate one of the filters to obtain a suitable brightness level; some have even reported an increase in image contrast with the polarizer in use.

Master solar observer Greg Piepol with the mounting of an adjustable iris diaphragm devised an alternative to the polarizing filter. Figure 3.3 illustrates his set-up with the blades fully opened and partially closed. Piepol notes that with the iris opening decreased the view darkens and allows subtle disc detail to be better seen.

Sunshades either mounted on the telescope or independently provide a means of avoiding eyestrain, fatigue, and possible sunburn. A telescope-mounted shade may be attached on either end of a refractor's tube. The Sun-facing side should be finished in white and the observer facing side a flat black. A freestanding solar shade is intended for the mirror-fed telescope not used in an enclosed room, but rather in the open air. Figure 3.4 shows amateur astronomer Robert Hess comfortably observing with his polar mounted solar telescope. The sunshade and Sun is positioned in this case conveniently at his back.

Fig. 3.3 Adjustable iris diaphragm for brightness control (Courtesy of Greg Piepol)

Fig. 3.4 Two possible routes for creating a sunshade for the solar observer. When mounting to a telescope, as on the *left*, make the shade's radius four times the objective diameter. Jamey Jenkins. On the *right* a freestanding shade is blocking the Sun's rays at the observer's back (Courtesy of Robert Hess)

Along the lines of the sunshade consider using a cloth, hood, or even a simple bath towel draped loosely over the head and eyepiece to provide a near-dark observing situation. You will particularly find that Ca-K observing is enhanced if the outside light is excluded; Ca-K views are inherently dark, coupled with the fact that your eyes may not be expressly sensitive to the wavelength. We have a tendency with age to loose the ability to deeply see in the violet.

Lastly, for a visual observer a means of measurement at the eyepiece is suggested. A filar micrometer, a graduated reticule, or a simple cross hair eyepiece will suffice. These accessories make it possible to determine the approximate size of sunspots or even prominence dimensions and their height above the photosphere. Combine a stopwatch with either of these accessories and after a little paper and pencil math you will have a completely new perspective on the vastness of solar features.

Finding Your Way Around the Sun

Most studies of the Sun require at least a rudimentary understanding of solar directions and movements. Defining directions on the Sun is not a difficult task. There are a few simple instructions to remember and with a bit of common sense, you'll immediately know your way around the Sun.

Solar Directions

Picture the Sun as it is crossing the meridian of your sky, that imaginary line passing from south, through the zenith overhead, to the north. Simply and approximately, the celestial directions on the Sun would then be as follows: the northern hemisphere of the Sun is toward our north celestial pole; the southern hemisphere of the Sun is toward the south celestial pole; the eastern solar hemisphere is toward Earth's eastern horizon; the western solar hemisphere is toward Earth's western horizon.

With the naked eye this is easy to comprehend. Looking through the eyepiece of a telescope that flips the view will be a bit more complicated.

Through any stationary telescope, the celestial western limb of the Sun will drift out of your field of view first, and as the Sun rotates, features disappear behind the western solar limb. Therefore, it can be said that the celestial eastern limb will drift out of the field last, while solar features rotate from around the eastern solar limb. The Sun turns on its axis from solar east to solar west. North and south celestial directions on the Sun can be easily determined through a telescope by simply nudging the instrument either north or south and noting the Sun's movement. For example, tap the telescope northward and the southern hemisphere of the Sun will want to leave the field of view and north will want to enter. Some observations will require no more than knowing these kinds of approximate *celestial* directions.

Because of the inclination of the Sun to the ecliptic and the tilt of the Earth's axis, the Sun's true north and south poles as viewed from Earth appear to tip and nod in the sky throughout the year. The Sun's *actual* north or south pole (axis of rotation) could, depending on the time of year, point toward or away from us (nod) by up to 7° 15′, and/or tilt to the east or west (tip) by as much as 26° 21′. An ephemeris like the one located at the back of this book is an excellent source for finding the tip and nod daily amounts (Fig. 3.5).

Solar Rotations

The outer layers of the Sun are not solid; they are gaseous plasma that rotates differentially, as if they were liquid. Consequently, those regions near the poles lag behind the more rapidly spinning equatorial zone of the Sun. The Sun rotates in the same direction as the Earth.

Two types of rotation periods exist, the synodic and the sidereal. Sidereal rotation is measured from a fixed point in space giving a period of 25.38 days for heliographic latitude of ±17°. A feature located at ±30° solar latitude takes about 27 days to complete one sidereal rotation, and the polar regions of the Sun, in excess of 30 days.

The synodic rotation period differs in that it takes into account the movement of the Earth in its orbit around the Sun. The *mean* synodic period is 27.28 days, a quantity that varies throughout the year because of the eccentricity of Earth's orbit.

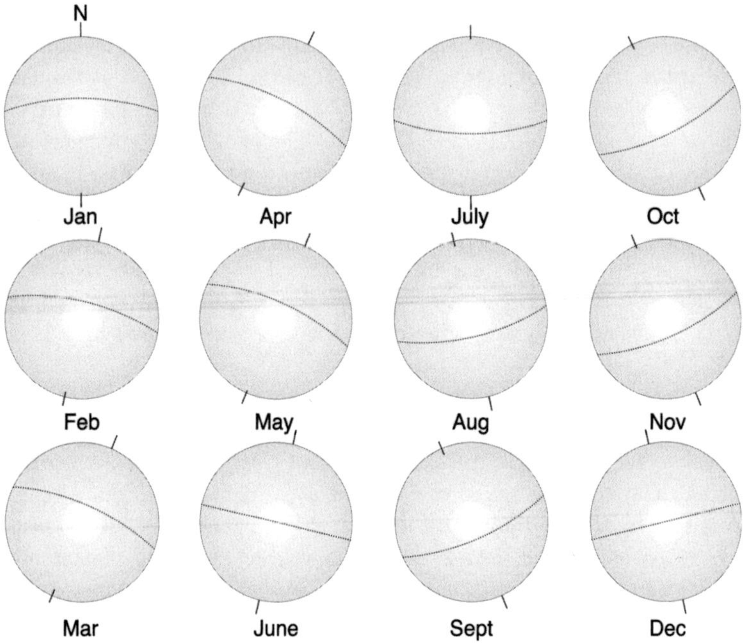

Fig. 3.5 The Sun's axes of orientation as seen in the sky vary throughout the year as illustrated in these drawings. North and south polar rotation points are indicated with the *solid lines*, north is up on each disc. The equator is shown with the *dotted line* crossing the mid-region of the globe. Note the months April and October, which exhibit the maximum tilt (called position angle) and the months June and December that have the least amount of nodding as we look squarely at the center of the Sun's disc

Astronomers have devised a system of numbering each rotation of the Sun. This is a method of defining solar time periods and keeping track of events happening on the Sun. Rotation periods are based on Richard Carrington's observations from the 1850s, at Greenwich Observatory. The so-called Carrington number identifies each rotation as a mean period of 27.28 days, each period beginning when 0° of solar longitude crosses the Sun's central meridian as seen from Earth. Carrington rotation numbers for each day can be found in astronomical almanacs published yearly or through various on-line sources. See the Appendix section of this book.

Active Region Numbering

When the Sun contains a localized event such as a sunspot, plage, or flare it is designated as an active region (abbreviation AR). Much of what solar astronomers observe is associated with active regions.

In order to maintain a record keeping system of solar activity, professional astronomers came up with a numbering plan that went into effect on January 5, 1972. Since that time, as each area of activity has been detected, that region has been assigned a sequential four-digit number (i.e. 4530, 4531, 4532...). The National Oceanic and Atmospheric Administration (NOAA) is given the responsibility of assigning the number to each active region. Some regions are long-lived and reappear at the east limb of the Sun repeatedly; these old regions are assigned a *new* number each time they appear.

Eventually all four-digit AR numbers would be used up and this indeed occurred on June 14, 2002 when AR10000 was observed. The solution since that time has been to retain the four-digit numbering sequence and ignore the repeating fifth digit. For example, AR10165 is simply referred to as AR0165.

Current AR numbers for solar features can be found over the Internet from a variety of sources. Several popular sites would include the NOAA, Spaceweather, and Mees Solar Observatory, University of Hawaii. The Mees site is an excellent resource for daily active region maps showing the location, size and AR number of each active region.

CHAPTER 4

RECORDING OBSERVATIONS

Amateur solar observing tends to fall into these three broad avenues of pursuit: statistical work, the study of the changing appearance of the Sun, and lastly casual viewing for pleasure, such as at an eclipse or one of the rare transits of Venus or Mercury. It's not unheard of for an ambitious individual to be active in all three areas.

Statistical observations are the procurement of data used to create various indexes that measure solar activity; this could include sunspot counting, tabulation of the heliographic latitude and longitude of activity centers, or perhaps in monochromatic light the determination of a monthly prominence number. To be effective these types of projects require an observer to examine the Sun as often as possible, such a commitment leaves a minimal number of gaps in the database.

Morphology is the term used to describe the study of the changing appearance of a celestial body. Obtaining whole disc photographs or drawing the daily movement of sunspots describes a morphological study. Producing a video of flare or prominence activity would also fall into this category. There are a number of activities that the amateur solar observer can tackle when studying the Sun, it is a truly vibrant body.

And of course from time to time the casual observer can always find something exciting happening on the Sun. Daytime astronomy provides an around the clock opportunity to enjoy a hobby that customarily only occurs "when the Sun goes down."

Observational data, whether sketches, notes, or photographic images are obtained to record what the observer has experienced, to provide a database for scientific research or to share with others in the hope that knowledge will be increased. If your motivation includes the latter, there is in the Appendix of this book a list of solar organizations dedicated to assisting an amateur astronomer with dispersing observations. It is highly recommended

Jamey L. Jenkins, *Observing the Sun: A Pocket Field Guide*,
Astronomer's Pocket Field Guide, DOI 10.1007/978-1-4614-8015-0_4,

that you contact one or two of these groups, inquire about their observing programs, and become an active participant. It is an excellent way to gain knowledge of the Sun, contribute to science, and make new acquaintances.

Counting Sunspots

For a solar observer seeking a white light activity that requires only a nominal amount of equipment, sunspot counting is by far the simplest way to reach that goal. A telescope fixed for projection or with an objective filter, a note pad, pencil, and a timepiece are all that is required for scientifically useful data gathering. The time spent at the telescope is fruitful, and requires only several minutes per day to complete an observation.

History

The existence of sunspots has been known since early Chinese astronomers first glimpsed large naked-eye examples many hundreds of years ago. It was not until the telescope first appeared on the scene about 1610 that the nature of sunspots became a bit clearer, revealing in addition the rotation of the Sun. Galileo and Scheiner were major contributors in this field. Heinrich Schwabe, through his search for a planet orbiting between the Sun and Mercury, stumbled upon the periodicity of sunspots. Schwabe recorded sunspots only to be able to discount them, in his hope of discovering a new planet.

The actual statistical study of the sunspot phenomena began around 1848, with observations conducted by the Swiss astronomer, Rudolph Wolf. Wolf was searching for a superior method of measuring solar activity. His initial choice, rather than counting sunspots, was to determine the weight of activity by area, but to do so he lacked equipment and expertise. Therefore, Wolf devised a scheme of counting the number of individual sunspots visible on the face of the Sun and then adding the number of groups seen multiplied by ten. His logic behind the totaling of the two quantities was that independently neither quantity alone was an accurate gauge of the true solar index, but combined they formed a reliable picture of solar activity. His theory was later confirmed by decades of observational statistics.

Obtaining Data

The official sunspot number is an estimate, a figure subjective to observations made by a number of independent observers. Several experienced amateurs have asserted that sunspot counting is more of an art than a science. This can be true, but only because of the factors that contribute to the difference in counts, such as the observer's experience, local seeing conditions, equipment, and the varying amount of solar activity within a 24-h observing window.

For instance, Wolf used an 80 mm refractor at a magnification of 64X for his sunspot counts. An observer possessing a larger aperture telescope, using increased magnification during similar seeing conditions could see less significant spots because of his telescope's greater resolution. And then there is the issue of what is a sunspot, and what is a pore? The sunspot counter is not expected to count a pore, but one person's interpretation of what is a pore may be different from another person's interpretation. Couple this with the fact that during the 24-h observing window, the hypothetical pore could mature into a spot or decay from view. All this meaning that your tally verses another observer's result depends on when they were made. As you can see, sunspot counting really is an estimate obtained by the work of an artisan, practicing a science.

The relative sunspot number, R the standard established by Wolf over 150 years ago, is computed from the formula $R=10\,g+s$. The total number of sunspot groups visible to an observer during a session is designated "g"; the total number of counted sunspots is "s." Computation is simple once the data has been obtained; often it is kept in tabular form in the observer's notebook. A scaling factor, called "k," is administered to an observer's submitted sunspot counts by the compiling organization. The purpose of k is to correct for the variables outlined above, that is, an observer's level of experience, local seeing conditions, and instrumental differences. Another purpose of k is to fine-tune modern day observations so a link is established with those from Wolf's time; that way continuity is maintained. The corrected relative sunspot number is therefore computed as $R=k(10\,g+s)$.

Sessions will vary amid solar observers depending on personal habits and observing styles. Several points must however be considered by anyone thinking of becoming a regular sunspot counter. First of all, regular observing is important for polishing the skills necessary for this activity. Generating a sunspot count on only 2–3 days a month guarantees poor results over the long term. It is required for a sunspot counter to obtain data as often as possible, ideally every clear day. Many observers try to find a convenient time for making the daily count.

Secondly, it is strongly recommended that a repeatable routine be instituted that allows stability from one session to the next. By being consistent in the observing plan, the likelihood of errors creeping in is lessened.

Here is a routine that one successful observer has performed for a number of years; your own plan may be different and reflect personal preferences. Begin by directing the telescope towards the Sun, either by watching the shadow formed on the ground or by utilizing a pinhole solar finder. If a projection screen is used the screen must be shaded from sunlight, otherwise small, dim sunspots will be difficult or impossible to see. For direct viewing a visual objective filter is favored. Since the majority of observers make use of direct viewing, the discussion from here on will focus on that method.

The sunspot count at the telescope is recorded on a pre-printed card, 100 × 150 mm (4 × 6 in.) in size and containing a disc for sketching the approximate position of all sunspot groups on the Sun. Space is provided for information regarding instrumentation, seeing conditions, and other pertinent notes. The sketch is useful when locating a sunspot group at the next observing session that may have evolved in a way that makes it difficult to recognize. Solar rotation, as you recall, will take all features a bit farther to the west.

In Fig. 4.1, a hypothetical example is given from 21 October 2012. The observer's refracting telescope uses an objective filter and a star diagonal giving the orientation seen on the drawing, north is up and west to the left. Your telescope may exhibit a different orientation. These are celestial directions and can be found by gently nudging the telescope N, S, E, or W and noting the displacement of the Sun. Select an eyepiece that that yields a magnification suitable for viewing the entire solar disc. For this stage if the eyepiece has an internal crosshair, this is all the better for determining the

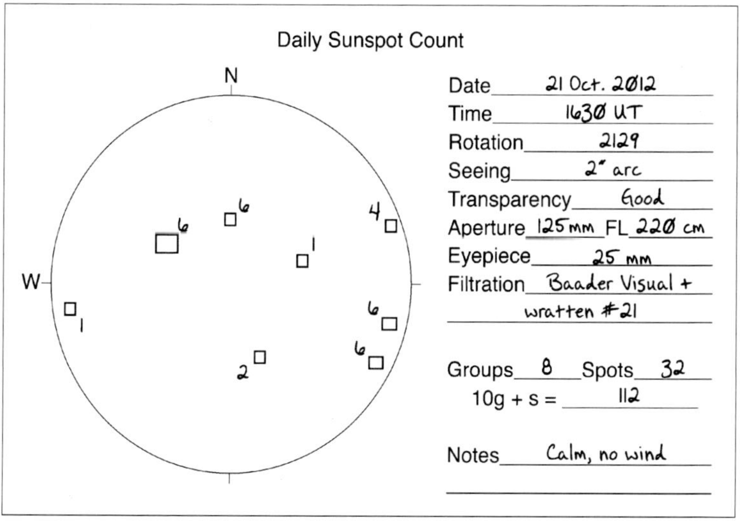

Fig. 4.1 File card used for recording sunspot counts

approximate location of sunspot groups and sketching them on the file card. If not, group locations can simply be estimated and drawn roughly in place.

The addition of a light red/orange filter on the eyepiece will serve to darken the umbra of a sunspot, making tiny, faint spots stand out from the background photosphere. Carefully focus the telescope on a sunspot group. If no group is immediately apparent, use the limb of the Sun. Consistency is important with any statistical program, therefore never bounce back and forth between different telescopes in this activity. Always use the same instrument and only log information that you are absolutely certain of.

While scanning inside the sunspot zones, to about 35° on either side of the solar equator, note all the groups visible, and sketch their locations on the card. Use a square or rectangle approximating the area of each group for this task. It may be necessary at times to "shake" the telescope slightly, to cause a small spot to be noticeable.

This is where the science becomes an art in solar observing. Familiarity with the McIntosh classification system is integral for recognizing what is an individual group and what may be two or more groups in close vicinity. Misinterpreting two groups as one will skew a daily sunspot count by 10, and if this occurs in several instances the count will be hopelessly inaccurate.

Having a card at the telescope with the McIntosh classes printed on it for reference is an excellent idea. If the pre-printed card system of recording observations appeals to you, listing the sunspot classes on the reverse side of the card would be advantageous. Blank copies of the card front and McIntosh classes for the back are provided in the Appendix section of this book.

Some observers use the "10-degree rule" when determining one group from another. This rule states "any spot or cluster of spots that is at least 10° of heliographic latitude or longitude away from any other spot counts as one sunspot group." Occasionally, two groups do form within 10° of each other, and the rule has to be modified to accommodate the situation. In most cases, however, the rule will apply and groups will be clearly separated by well over 10°.

Regular solar observing is an aid to determining group status. Why? Regular observing allows the astronomer to witness a group's evolutionary progress. There are days when an observer will not see the Sun, and on those days consider using the Internet to maintain a connection with solar activity. However, refrain from "checking out" the Sun online if plans are made to observe later that day. That way the daily sunspot count will not be influenced by what may be seen on the web.

Having sketched the sunspot groups that are seen at a whole disc magnification, the eyepiece is then switched to one of 90–100 power, and beginning at one limb carefully scan the sunspot zones, searching for any additional tiny, compact, or faint groups. Remember that a single isolated spot is considered a group. If one is swept up, include it on your sketch. Again use a red/orange supplementary filter to help distinguish a spot from the background; a lighter shade than before may be necessary because a greater magnification results in a dimmer view.

Once confident that all groups have been recorded, and still using the 90–100 power eyepiece, return to one side of the Sun and inspect each group, counting individual sunspots. The total number of spots in each group is written beside its corresponding square or rectangle on the card. Move from one group to the next, until all the spots have been counted.

The inclination at this point is to ask, "What is a spot and what is not?" A granular feature can appear dull; a pore is darker than a granule with an average diameter of 2–3 arc sec; neither of which counts as a sunspot. A sunspot umbra (which is what is counted) is darker than a pore and may be several arc seconds in diameter and up to any size large enough to be glimpsed with the naked eye. Any spot that is conjoined is counted as a single spot, as soon as it separates and the pieces of umbra no longer touch, the pieces become individual sunspots. If a light bridge is dividing a sunspot, consider it a single sunspot until the spot is completely divided. Penumbrae and pieces of penumbra are not counted as sunspots. If it is difficult to distinguish small pieces of penumbrae from umbra, use the brightness test by comparing the suspect spot with nearby known umbra. If the brightness difference is significant, consider the suspect to be penumbra. If similar, count it as a sunspot. It is only with years of observing that we learn experience to be the best teacher in the "art" of sunspot counting.

After all the spots have been counted, remember that during sunspot maximum the total can easily be over a 100, return to the starting point and repeat the count. It is easy to miss a spot here or there, so correct the file card as needed. After the second confirming spot count, fill in the remaining data on the card: the date, the Universal Time when the count was completed, the Carrington rotation (found in the Appendix at the back of this book), an estimated interpretation of seeing and transparency conditions and data regarding instruments used. Note that the eyepiece identified is the one used to make the spot count, not the initial group evaluation. The relative sunspot number is calculated from the group and sunspot totals and entered on the card. The cards from all observations are filed until the monthly report form is due to be completed and submitted to the appropriate data gathering organization.

Daily Sunspot Count MONTH/YEAR June / 2012

	UT	S	T	Gn	Sn	Gs	Ss	G-total	S-total	R
01	1530	2"	G	2	12	2	12	4	24	64
02	1530	2"	G	2	15	1	6	3	21	51
03										
04										
05	1545	<1"	E	4	25	2	2	6	27	87
06	1540	2"	G	4	50	2	6	6	56	116
07										
08										
09										
10	1615	5"	F	3	87	3	59	6	146	206
11	1619	2"	F	4	83	3	69	7	152	222
12	1544	1"	G	3	51	3	89	6	140	200
13	1530	1"	G	3	30	3	85	6	115	175
14	1530	2"	G	3	9	3	42	6	51	111
15										
16										
17	1650	3"	G	2	14	4	29	6	43	103
18	1647	4"	G	2	15	4	26	6	41	101
19	1550	4"	F	2	17	4	57	6	74	134
20										
21	1626	2"	F	3	26	3	72	6	98	158
22	1645	2"	F	2	20	2	54	4	74	114
23	1640	3"	F	2	33	3	31	5	64	114
24	1635	2"	F	3	41	2	35	5	76	126
25	1700	1"	G	5	38	3	16	8	54	134
26	1725	1"	G	6	54	2	14	8	68	148
27										
28										
29	1538	3"	G	7	32	3	33	10	65	165
30										
31										
Monthly Total				62	652	52	737	114	1389	2529
Monthly Average				3.3	34.3	2.7	38.8	6	73.1	133

Aperture 125mm fl 2200mm Eyepiece fl 25mm Magnification 90x

Filtration Baader Visual Objective + Wratten #21

Fig. 4.2 Recording log sheet for the month of June 2012

An alternative to the file card method of recording, one intended for a statistician wishing to have a more detailed program of sunspot counting is to maintain a log sheet. A completed sample log sheet is provided in Fig. 4.2 illustrating a typical month of sunspot counting. The monthly submission form provided by several national organizations is similar in format to this one; it is possible on a home computer to design your own form, printing copies as needed. You will immediately notice that additional information is requested on the log sheet that doesn't appear on a daily card.

Across the head of the page, columns are identified as follows:

- UT is the Universal Time at the completion of the observation
- S is the seeing conditions defined by the arc-second scale
- T is the transparency of the sky: excellent, good, fair, or poor
- Gn is the number of sunspot groups north of the solar equator
- Sn is the number of sunspots north of the solar equator
- Gs is the number of sunspot groups south of the solar equator
- Ss is the number of sunspots south of the solar equator
- G-total is the sum of Gn and Gs
- S-total is the sum of Sn and Ss
- R is the relative sunspot number, computed from $R = 10\,g + s$

The monthly column totals and averages are entered along the bottom for convenience. An average is based on the number of observation days, *not* the number of days in a month. The telescope data at the foot of the log sheet could be preprinted on each sheet since the instrumentation is to be used repeatedly. Note that the magnification sited is the one used for counting the individual sunspots, not the groups.

It is possible to glean *R* from the cards or a log sheet and graphically present it against time as is illustrated in Fig. 2.5. Also available from either of these tools is a sunspot group statistic called the mean daily frequency (MDF). To calculate the MDF for the month, add the daily totals of sunspot groups and divide this by the number of days observed. In Fig. 4.2 the MDF for the month is 6, calculated as the sum of the column "G-total" divided by 19, the number of days with entries. When diagramed against time, the MDF will also show the solar activity cycle.

An even more complex study can be conducted by tabulating groups and spots based on the hemisphere in which they develop. The log sheet organizes this information in the columns Gn, Sn, Gs, and Ss. Collecting this data is important because the amount of activity occurring in one hemisphere can be substantially different than that occurring in the other. When the latitude of sunspot groups is displayed against time a unique pattern called the butterfly diagram develops, see Fig. 2.4. A butterfly diagram visualizes the migration of sunspots from the higher latitudes toward the solar equator as a solar activity cycle advances.

The key to sorting out these detailed studies is recognizing whether a group is north or south of the solar equator. As you recall, the Sun appears to nod and tip as the Earth swings through its yearly orbit, changing the apparent position of the solar equator from month to month. The following pages contain instructions on determining the position of the solar equator.

Recording Sunspot Positions

An informative activity many solar observers and several professional observatories practice is the daily recording of sunspot positions. Useful for interpreting the motions of sunspots as well as picturing the rotation of the Sun, this interesting venture is accomplished with handcrafted drawings. Detailed here you will find a simple, yet precise method that utilizes no more than a sharp No. 2 pencil, a sheet of paper, and a solar projection screen.

The basic technique is to trace the sunspot locations inside a circular area on a prepared form containing the projected image of the Sun. On the same form is drawn a reference line marking accurately East–west or North–south. This elemental diagram, transferred to a standard recording form includes the necessary information for locating the solar equator, and from there finding heliographic coordinates of individual features.

To begin, a clock driven telescope fitted with a projection screen is set up to follow the Sun. A non-cemented eyepiece that projects a solar image having a diameter of 150–180 mm is ideal. This size is large enough for good detail to be visible, but not so large that the projected image becomes dim and difficult to see. Make sure the projection distance is adjustable to allow for the varying angular diameter of the Sun. Because the desire is to establish a uniform size to the projected disc; it could be worthwhile to have a circle of the correct diameter drawn on the screen. Use this circle to pinpoint the projection distance at each observing session. The telescope and projection screen mounts must be sturdy otherwise potential vibration from wind or the application of pencil marks to the drawing paper may disturb the telescope's pointing.

Determining N-S and E-W

Once the Sun is accurately sized and focused on the projection screen, attach a drawing blank; see Fig. 4.3, to the screen. It will be advantageous later if the blank is of a lightweight or translucent paper. Spring clips can

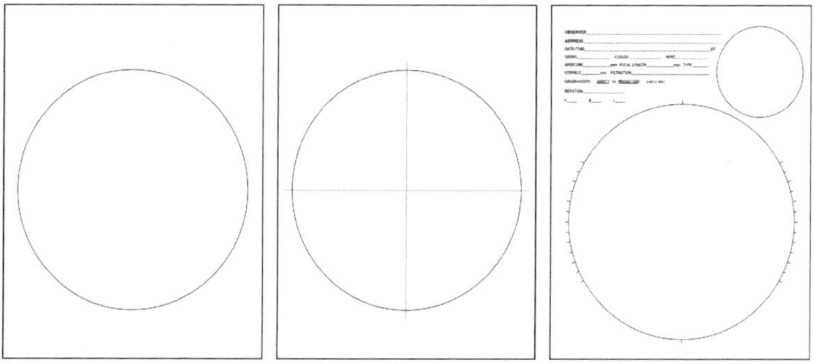

Fig. 4.3 Three variations on the drawing blank for recording sunspot positions include the plain circle, a circle with bisecting lines, or a standard recording form, this one adapted from the Association of Lunar and Planetary Observers Solar Section. The larger disc diameter should be 150–180 mm

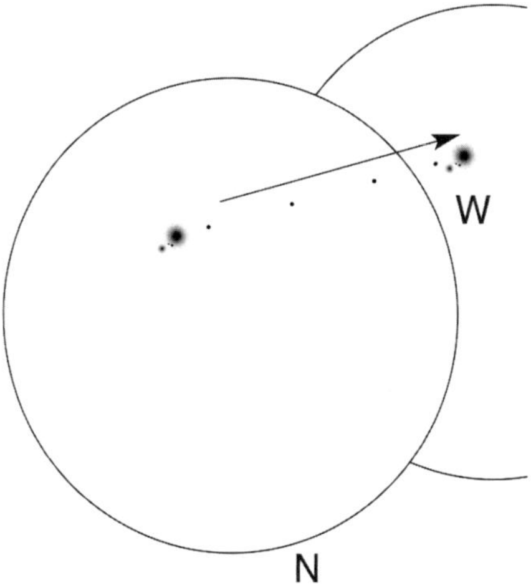

Fig. 4.4 Determining celestial directions using the drift method. Dots representing the position of a sunspot's umbra are drawn on the blank form as the rotation of the Earth drifts the Sun to the west. A line connecting the points locates E-W; a line perpendicular to E-W determines N-S

be used to hold it in place. The content of the blank is to be a plain circle, a circle with bisecting lines, or even the actual recording document provided by a national organization that archives this data. As the technique is refined, you'll decide which option of the drawing blank works best.

Center the Sun within the circle on the drawing blank, this will be made easier if the mounting has electric slow motion control. Locate a tiny sunspot and carefully mark its position with the pencil using a small dot in the center of the umbra. Switch off the telescope drive and watch as the Sun and the sunspot slowly drift off the circle. The projected image is drifting toward celestial west because of the Earth's rotation. As the spot drifts from its original position carefully pencil a small dot centered on the umbra each time it moves about 25 mm. After a minute or so, when the sunspot has drifted a distance beyond the circle, draw a final dot and switch the telescope drive back on. An example of such a blank is provided in Fig. 4.4. Re-center the Sun on the drawing blank.

The series of dots you have just drawn define a line representing the direction E-W as projected onto the celestial sphere. Strike another line perpendicular to this one and the N-S celestial direction is indicated. Knowing these directions is essential when using a grid overlay to determine heliographic latitude and longitude on the Sun. Mark on the drawing an "E" indicating the east limb of the Sun; this is the side opposite the direction of drift. Using the telescope's declination slow motion or by gently nudging, shift the telescope to the north and note which limb of the Sun moves toward the center of the projection screen. This is the north limb; mark it with an "N" on the drawing blank.

If the drawing blank is prepared with a set of bisecting lines (to indicate the celestial directions) rotate the blank form about the optical axis of the telescope until a sunspot drifts accurately along one of the lines. That line will be E-W the other is N-S.

There may be a time when finding E-W using the previous drift method is not convenient. On those occasions the N-S direction can be located using this technique. Accurately center the projected Sun on the drawing blank. Switch off the electric drive and watch as the Sun drifts toward the west. When the drifted Sun has overlapped the drawing blank as in Fig. 4.5, switch the driving mechanism back on and then quickly and carefully draw two marks at the points of intersection, X and Y. A line connecting the two points indicates the N-S direction. E-W is perpendicular to that line. Whichever method you chose, after the directions are determined, the drawing blank is now ready for recording sunspots.

Tip

For determining the celestial E-W orientation accurately in a photograph the drift method is also applicable. Simply observe the drift of a notable feature through the camera's viewing system and rotate the camera around the telescope's optical axis until the feature tracks exactly along the edge of the image frame. For later reference during image processing, at the telescope note which sides of the image frame are east and north.

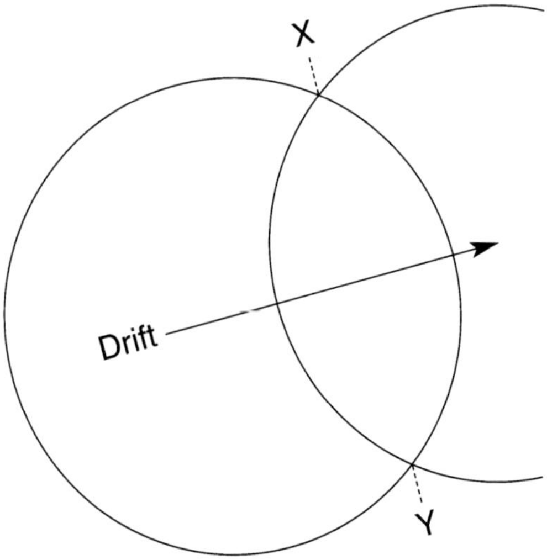

Fig. 4.5 Locating N-S can be accomplished by drifting the Sun until the east limb intersects the west. Quickly draw a mark at each of the two intersecting points (X/Y). A line drawn between X and Y locates the N-S celestial direction

Drawing Sunspots

The key to drawing sunspots for a position measurement is to work quickly. A solar feature is constantly in a state of evolution and therefore potentially changes appearance relatively rapidly.

Begin the drawing process by first marking the position of all large spots to be drawn with a dot or hash mark. Then return to the first group only penciling in the umbrae, pay particular attention to the size and shape of the features. Include smaller attending sunspots, and then move to the next group. After all sunspot umbrae are sketched return to the starting point and trace the penumbra. The penumbra is shaped by forming a line around the periphery of the umbra. Do not attempt to shade penumbra, simply outline the outer boundary of the penumbral areas.

Faculae may be visible and included in the white light drawing. Use a dashed outline to indicate a facular region, illustrating the difference between penumbrae and faculae. If the observer decides to include a facular

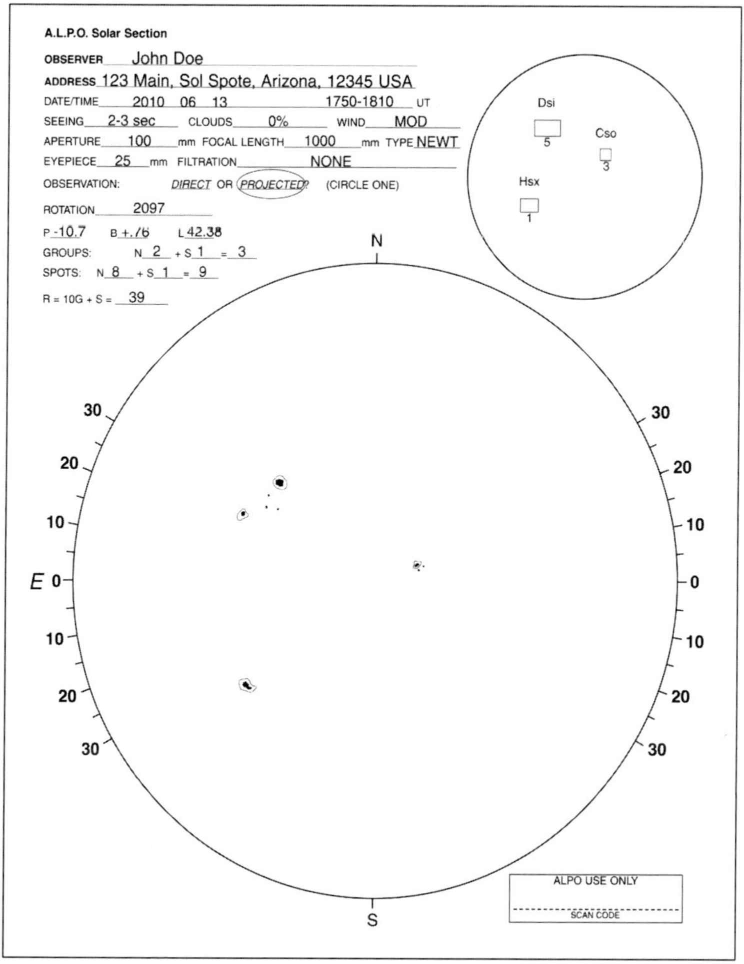

Fig. 4.6 A completed whole disc sunspot drawing of the Sun. Faculae have not been included in this view

region in their drawing they ought to make an estimate of the feature's brightness in some standardized manner. A three-step scale is sufficient for this purpose. **Dim** is to be used for any region that is only slightly brighter than the normal photosphere. **Bright** is intended for facula that is fairly seen near the limb. **Intense** is reserved for any example that is very easily seen, especially nearer the center of the Sun's disc (Fig. 4.6).

At this point the rough sketch is completed and observing notes ought to be jotted down for inclusion on the finished drawing. These notes would include information about the telescope used: aperture, focal length, type. The seeing conditions are important as well as an estimation of cloud cover. Do not forget the date and the starting/ending Universal Time of the observation (most important!).

To make a proper presentation of your sketch, a standard report form should either be created or adopted from one of the national solar observing organizations (i.e. ALPO, BAA). This form will provide spaces for all the relative data concerning an observation along with a suitable sized disc (same diameter as the drawing blank) for the artwork.

The often used orientation for a photograph or drawing of the Sun is *celestial* north - up and east - left, just as the Sun located on the northern hemisphere meridian appears to a naked-eye observer. A rough drawing most likely will not have this orientation because of the optical configuration of the telescope, which may flip and/or flop the image. For convenience therefore many observers choose to use the orientation north - up and west - left as would be created with a refracting telescope projecting directly behind the objective. Any orientation is acceptable as long as celestial directions are indicated.

Parallel to the drift marks on the drawing create a line that bisects the circle. Then to that draw another bisecting perpendicular line, label the north position and the east or west position. The bisecting lines do not represent the pole region or equator of the Sun, those will be located later. What they do represent are the cardinal celestial directions, as seen from Earth. The task now is to transfer the data from the drawing blank to the presentation reporting form. A small light box is useful to allow any required flipping of the *original* drawing to the desired orientation. Align the recording form *over* the drawing blank on the light table and register the disc circles while adjusting placement of the N-S/E-W tic marks over the N-S/E-W lines of the drawing blank. Trace the umbrae, penumbrae and faculae onto the recording form. Finally, complete the data entry on the recording form as is pertinent to the observation.

Determining Heliographic Coordinates

This scheme is used for locating the equator or heliographic coordinates on a whole disc sunspot drawing/photograph that has accurate markings of celestial N-S/E-W.

Because the orbital plane of our planet is a bit out of sync with the solar equator, the lines of latitude and longitude on the Sun appear to nod and tilt as seen from our Earthly perspective. Determining the latitude and longitude therefore require that we recognize how much the Sun is tilting and nodding the day of the observation (see Chap. 3, Finding Your Way Around the Sun).

There are three elements needed from the date/time of the observation that are used to find the heliographic coordinates of a feature on the solar disc. These elements *Bo*, *P*, and *Lo* can be obtained from the Ephemeris located in the Appendix of this book, or from published tables found online or in book form at a public library.

The first element is the amount of "nod" that the solar poles are showing at the time of the observation. This is termed *Bo* and actually denotes the changing heliographic latitude at the center of the solar disc. The maximum nod of the Sun's northern hemisphere toward Earth, +7.3°, occurs early in the month of September. Conversely, during the first week of March, the southern pole region of the Sun is utmost visible, as the latitude at the center of the disc becomes −7.3°. There prevails only a short time each year when *Bo* is 0° and we are looking squarely at the center of the Sun's disc; these brief periods are found in June and December.

The second element is designated *P* for position angle. *P* represents the amount of tilt the north solar axis has relative to the rotational axis of the Earth. The magnitude of this offset or tilt varies throughout the year by a total of 52.6°. When the north solar pole is tipping to the east of celestial north, *P* values are given a positive (+) sign. The maximum tilt occurs in the early days of October, with a value of +26.3°. As we progress in our yearly orbit, the Sun begins tipping back to the west until its N-S axis is aligned with celestial N-S in early January, at that time $P = 0°$. Beyond this date the values of *P* begin to assume a negative (−) sign, reaching a maximum westerly tilt of −26.3° around the beginning of April. The tilting motion then reverses, with the north pole edging back toward the east, passing through celestial north again, and a $P=0°$ near the first week of July.

These two elements, *Bo* and *P* from the date of an observation are all that is necessary to accurately fix the solar equator or the latitude of a feature. To find the longitude requires the third element, an imaginary line drawn from north to south representing the Sun's axis of rotation. This line is called the central meridian (CM) of the Sun. A solar astronomer calculates a feature's longitude by its distance from the CM at the date/time of the observation.

The CM longitude (abbreviated *Lo*) is 360° at the beginning of each new solar rotation. Heliographic longitude is measured increasingly from east to west, and because the Sun rotates from east to west the longitude at the CM decreases with time from 360° to 359° to 358°, and so on until a solar rotation completes. *Lo* experiences a daily decrease of about 13.2°. This converts to an approximate hourly rate of 0.55°. The same Ephemeris that provides the *Bo* and *P* values is used to find the nearest value of *Lo* to the observation time. *Lo* is normally provided in the daily table for 0-h UT. That information is then interpolated using the above knowledge to obtain at the time of the observation, the longitude of the CM. Once the longitude of the CM at the observing time is known, it is a simple matter of measuring the distance between a feature and the CM to conclude the feature's longitude.

A whole disc drawing or photograph is easily evaluated with a latitude and longitude accuracy of about 1° near the center of the disc, and because of foreshortening with a slightly less exactness approaching the limb. Mathematically an observer can accomplish this task, however many prefer the far less tedious method outlined here which utilizes an overlay system of Stonyhurst Discs (Fig. 4.7).

A Stonyhurst Disc is a template showing the lines of latitude and longitude for a specific value of *Bo*. The typical set consists of eight individual overlays created with *Bo* increments from 0° through +7°. By flipping the grids head for foot, the opposing *Bo* values of −1° to −7° become available. Stonyhurst Discs are obtainable through some observing organizations, but the easiest way is to download them through an Internet source and

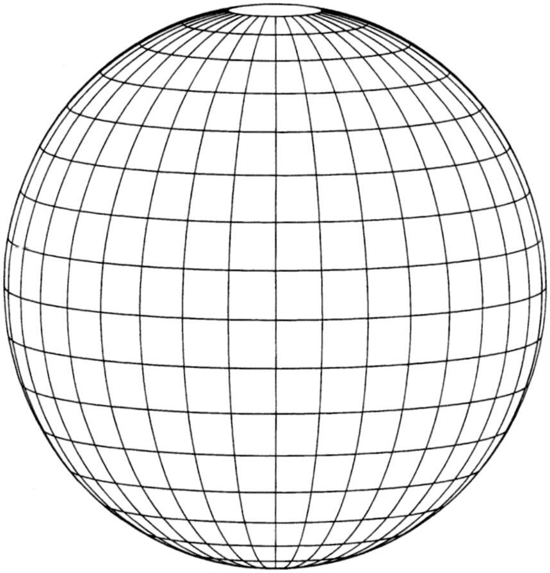

Fig. 4.7 Stonyhurst Disc with a *Bo* value of +7°. Flip the template head for foot, and the *Bo* becomes −7°

print discs from your computer. Several options exist including the resource provided by British amateur observer, Peter Meadows (www.petermeadows.com). His site provides links to discs in several formats in a variety of diameters to meet any amateur solar observer's needs. Meadows has also assembled excellent tutorial material and has available a freeware program called Helio. His program can be used to determine sunspot positions from X and Y coordinates on the solar disc, as well as the daily parameters for *P*, *Bo*, and *Lo*. Additionally, a set of Stonyhurst Discs is provided in this book's Appendix section, scaled copies can be created for your personal use.

The whole disc drawing of the Sun *must* be of the same diameter as the Stonyhurst Disc. Standard disc size for a drawing or photo is 150–180 mm. Because the overlay is to be placed above the drawing it could be difficult to see through a relatively opaque paper, therefore print the Stonyhurst Disc on a clear or translucent media.

The necessary tools to measure a drawing are: a Stonyhurst Disc of the proper *Bo*, a protractor, a ruler, a fine tip ink pen or sharp pencil, several paper clips or scotch tape, and the values of *P* and *Lo* for the day and time of observation. After a few drawings or photos have been measured you may well determine a shortcut or two that will save you time and steps.

To begin, the drawing must have the E-W and N-S celestial directions marked within about 1°, refer to Fig. 4.8. With these markings on the drawing, north is located at the top and east or west to the left, depending on the orientation you have selected. From the cardinal directions divide the Sun's disc into four quadrants; the purpose of this is to just locate the disc's center. Take the protractor and measure from the north celestial point on the Sun the amount of *P* given in the Ephemeris for the day of observation. Positive values are measured from celestial north toward the east and negative values toward the west. Draw tic marks on the ends of an imaginary line passing from the point of *P* through the center of the Sun and exiting the opposite hemisphere. The upper tic mark represents the position of the true north pole of the Sun and the lower mark the southern pole.

As illustrated by Fig. 4.9, a Stonyhurst Disc having the proper value of *Bo* is now positioned over the sunspot drawing. As the disc is orientated remember that a positive *Bo* causes the north polar region of the Sun to nod toward the observer, a negative value causes it to recede. When selecting the disc, pick the one with the nearest *Bo* value for the day of the observation. For instance, if the *Bo* from the Ephemeris is +4.6, select the +5 *Bo* Stonyhurst Disc. Center the disc over the drawing and rotate the CM longitude line until it intercepts the north and south tic marks of the *P* value drawn. Using paper clips or scotch tape temporarily affix the Stonyhurst Disc to the drawing.

The sandwich of drawing and overlay reveals the location of the lines of latitude and longitude on the photosphere. Finding the latitude of a feature is rather straightforward; the equator is designated as 0° and each line above or below represents a 10° increment. Some features may fall between the lines and their position will have to be interpolated; the ruler is helpful for this purpose. A latitude north of the equator is designated as N or (+), and any to the south are S or (–).

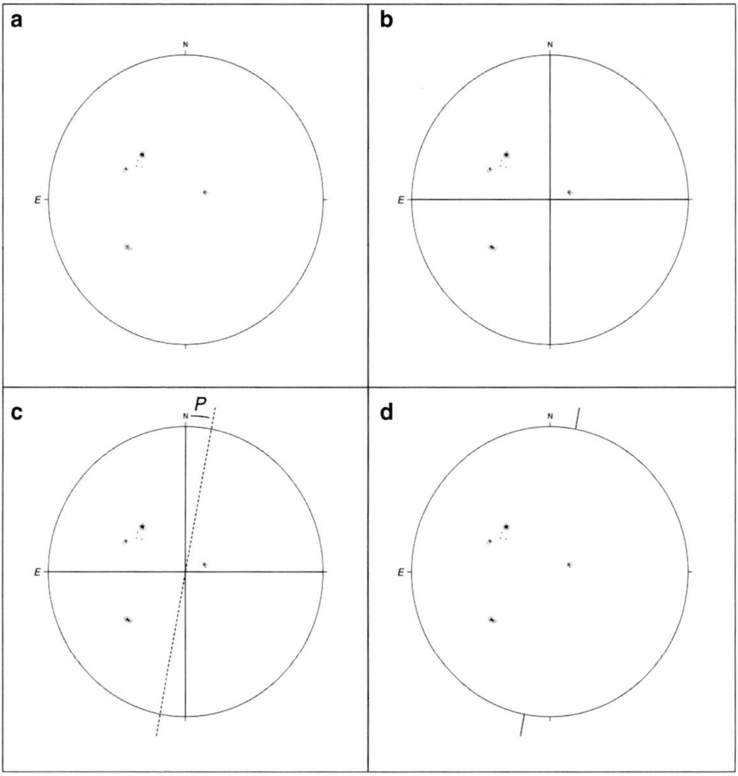

Fig. 4.8 How to find the north and south poles on the Sun. In this instance the illustration from Fig. 4.6 is used. Figure "a" is the sunspot drawing, which indicates the celestial directions N-S and E-W. First bisect the drawing as shown in figure "b" to locate the center of the disc. Next, as in "c", mark the number of *P* degrees from the north celestial point on the day of observation. From the ephemeris, *P* has a value of −10.7°. Negative values are located on the west side of celestial north. Draw tic marks at this angle representing the Sun's true north and south poles as seen in figure "d"

Solar longitude is measured from the central meridian and can be presented in either of two forms: the Carrington system or the more convenient "relative" system. With the Carrington system it is necessary to first calculate the longitude (*L*o) of the CM at the time of the observation and then measure the feature's distance from the CM. Once the distance in

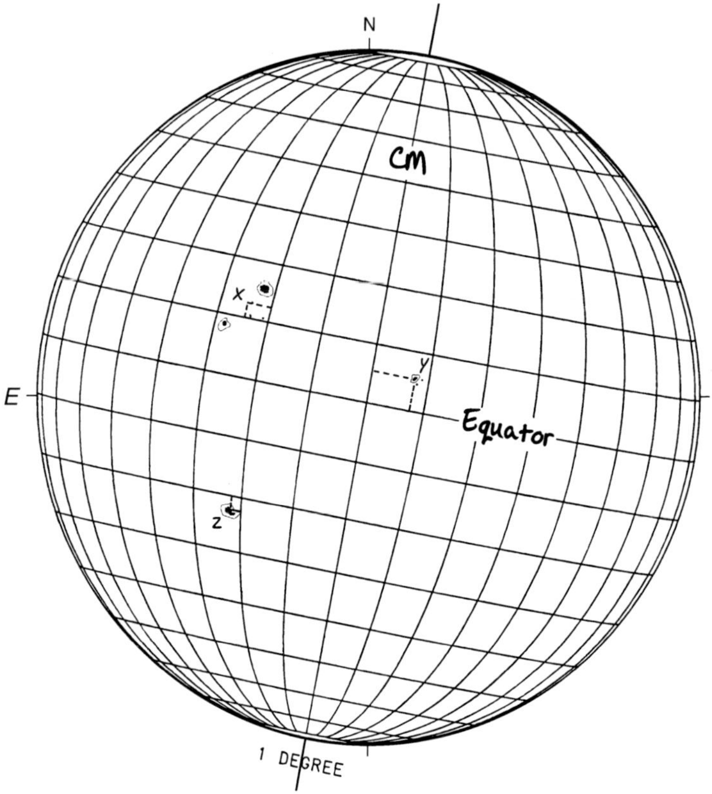

Fig. 4.9 Positioning the Stonyhurst Disc over the Fig. 4.6 drawing. For this illustration the correct Bo value is +0.76. The nearest Stonyhurst Disc from a conventional set of eight is the +1 disc; rotate the disc until the CM of the template intersects the tic marks drawn that represent the N-S poles of the Sun

degrees from the CM to the feature is known, add or subtract that distance from the CM longitude. Longitude to the west of the CM will be greater than the *Lo*; therefore add the feature's distance to obtain its longitude. A feature to the east of the CM is subtracted from the *Lo* to obtain its longitude. The Carrington system is used for precise investigation and reporting of active regions.

The relative longitude system simply states a feature's distance in degrees from the central meridian, whose reading is accepted to be 0° at the time of observation. Longitude east of the CM is designated E, and longitude to the west as W. Relative longitude is convenient and is widely accepted, so long as the date and time of the observation are also noted with the heliographic coordinates. Such relative observations are obtained daily with the Mt. Wilson Observatory 150-foot solar tower.

As an exercise, determine the *relative* heliographic coordinates of sunspot groups X, Y, and Z found in Fig. 4.9. Each division represents 10° of latitude or longitude, assume the longitude of the central meridian is 0°; therefore the intersection of the CM and the solar equator has coordinates of 0°, 0°. A millimeter rule is helpful when dividing the space between the lines, which varies because of foreshortening as you approach the solar limb. Fractions of a degree are normally rounded to the nearest digit so that the heliographic coordinates would read as follows:

- Group X = N14 E25
- Group Y = N6 W8
- Group Z = S23 W23

Of course with relative coordinates the date *must* be referenced, include the beginning and ending universal time of the observation. On the order of 2010.06.13.1750–1810 would be acceptable, following the nomenclature of YearMonthDayUT beginning-UT ending. The Carrington longitude of a feature can be obtained when the precise longitude of the CM is determined and the addition or subtraction of the relative longitude coordinate is administered to that figure.

Measuring Prominences

Why would an amateur solar observer be interested in determining the position of a prominence or the height of a hedgerow? Determining the position of a prominence is helpful for identification purposes as well as studying prom frequency with regard to specific latitude. Knowing the height of a prominence can be indicative of a soon to be seen eruption. Additionally for the amateur estimating the height or velocity of a prominence, simply satisfying a curiosity about the feature is what's important. It's educational and interesting to know that a feature under study is many times the size of Earth, and perhaps is being ejected from the Sun at a truly astronomical speed.

The position and size of a chromospheric disc feature in a whole disc photo can be found using the Stonyhurst methods outlined on the previous pages, if the photo is scaled to match the Stonyhurst Disc, and accurate indicators of N-S and E-W are included.

Position Angle

Measuring prominences as they occur at the solar limb is a relatively easy task. The features visible are nearly in profile, their alignment to the observer being roughly 90°. Position angle (P.A.) is a convenient means of identifying the location of features that frequently appear on the solar limb.

On the Sun, P.A. is determined from the north polar axis moving eastward (counterclockwise) and is divided into 360° around the limb. Position angle is read therefore as follows: N = 0°, E = 90°, S = 180° and W = 270°. To measure a photograph a clear cell or translucent template is prepared with a diameter equal to the standard disc size (150–180 mm); then serving as an overlay, the cell is sandwiched with the same sized chromospheric photograph. A master template for this work is provided in the Appendix.

Like the sunspot drawing, a sun photo must contain accurate indicators of the celestial directions N-S-E-W. If the template is printed black on a clear background consider producing a negative print of the Sun, which will make reading the P.A. divisions easier. Refer to Fig. 4.10. The template is to be positioned on the photo in accordance with the celestial directions, and tilted to the East or West by the amount of *P* indicated in the daily ephemeris. Remember that the true north axis is tipped to the east of celestial north when a *P* value is given a positive sign and to the west if the value is negative. The P.A. for the feature can now be read directly from the overlay using this method.

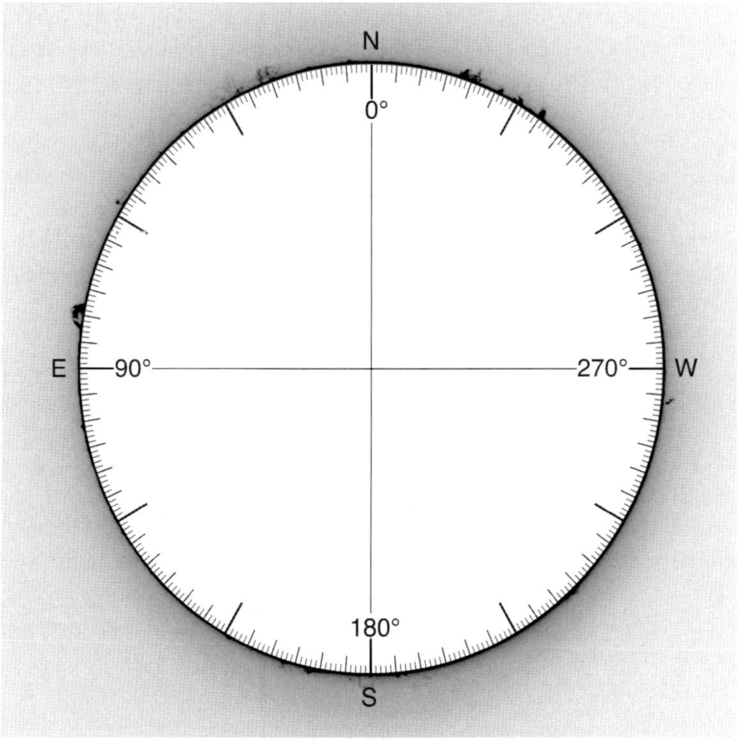

Fig. 4.10 Position angle is the measurement around the limb of the Sun beginning at north and progressing to the east. A few examples include the hedgerow formation located between position angles 17°–33°, the broken arch prominence at 80°, the leaning pillar at 263°, and the field of active proms located from 324°–343° (Negative print from an H-alpha image by Jim Ferreira, 31 July 2010)

A direct visual approach is to use a protractor type reticule with an eyepiece that permits viewing of the Sun's entire disc. A reticule of this type is built into a few astrophotography-guiding eyepieces or can be obtained through optical houses and surplus dealers. At the telescope, the reticule must first be orientated to celestial E-W by using the drift method. Once a thread is aligned accurately resume the driving motion of the mounting and slew the telescope so the monochromatically filtered Sun is centered on the protractor. Looking through the telescope, the P.A. of any prominences will be immediately evident.

At this point you might want to have a drawing blank prepared that has a lightly printed representation of the reticule on it surrounded by a disc portraying the limb. Instead of drawing the rapidly changing prominences quickly sketch squares or rectangles on the edge of the circle depicting the prom's position and size. Once all the prominence's positions have been sketched, start at the celestial north point labeling them numerically (1, 2, 3, etc.) through the east and around the limb returning to the starting point. This serves as an identification key if you create photographs or detailed drawings of an individual prom.

The P.A. is read directly from the drawing blank being certain to include correction for the daily *P* value. For example, if the P.A. of a prominence is shown to be 87° with the reticule and the *P* value on the observation day is −10°, the corrected P.A. of the prom is 97° (87 + 10). Negative *P* values cause the Sun's axis of rotation to tilt to the west thereby increasing the P.A. around the limb relative to celestial north. Positive *P* values indicate an easterly tilt and require an opposite adjustment or reduction from the celestial position angle. If you are careful a prominence position can be identified within a degree or so. Besides identification this method is suitable for anyone interested in studying the frequency of prominences at various solar latitudes.

Finding Prominence Height

Determining the height of a prominence above the solar limb from a whole disc photo is a process based on knowing the diameter of the solar disc. The solar disc has an accepted diameter of 1,391,980 km (radius 695,990 km). For comparison the Earth has a diameter of 12,756 km.

With a whole disc photo of the Sun (the larger the photo, the greater the accuracy) and a finely divided millimeter ruler, measure the diameter of the Sun and the height of the prominence above the limb. Divide the prominence measurement by the disc measurement and multiply the resulting quotient by the diameter of the Sun in kilometers to find the approximate height of the prominence in kilometers.

A visual method of determining prominence height requires the amateur to acquire either a graduated reticule or a filar micrometer. A timepiece is used to calibrate the tic marks of the reticule or the vernier scale of the micrometer. Because of the variable distance to the Sun throughout the year the calculations are performed each day a measurement is taken.

Begin by orienting the measurement scale in the celestial E-W direction by using the drift method, preferably with a magnification of 100x or greater. Once the scale is parallel to the drift, position the Sun in the telescope so that the celestial west point of the limb is tangent to a thread of the graduated scale, disengage the telescope drive and time the transit of the entire solar disc across this thread. Repeat the procedure several times and average the results.

Now return the Sun to the transit position as described above, and time the interval of the west limb drifting between two threads of the scale. Repeat this several more times and average the results. Divide the whole disc transit time by the drift time between the two threads to obtain the disc diameter to scale factor. Divide the known diameter of the Sun (1,391,980 km) by this factor to find the image scale, which is the number of kilometers, each division of the reticule or micrometer scale represents. So long as the diameter of the virtual image of the Sun in the telescope doesn't change with respect to the graduated marks, the image scale will remain the same. Once the reticule or vernier is calibrated, it's possible to position the scale over a prominence, count the number of divisions consumed, and multiply that number by the image scale to find the prom's height above the limb.

Velocity

Determining the velocity of an eruptive prominence requires obtaining a series of photographs that are accurately timed, and show a minimum of two points of reference located on or very near the solar limb.

Reference points could include two quiescent prominences or a quiescent prominence and a sunspot. Knots of gas within the erupting prom are carefully identified in each photo; these are then measured for their displacement relative to their previous position while noting the amount of time (in seconds) passed between successive photos. The scale of the photos (km/mm) must be known or the center of the Sun visible in each photo to determine the scale.

The basic technique can be demonstrated by the following procedure. Obtain a photo series that meets the above criteria and arrange them on a table before you. Use tracing paper over the first photo and mark with a fine point pen or pencil the reference points, the solar limb, and the center of the gas knots you will be measuring (if the scale is not known, also mark the position of the center of the solar disc). Transfer the tracing paper to the second photo of the series and carefully align the tracing to the same reference points; mark the new position of the gas knots of the eruptive prominence. Repeat this procedure until the last photo has been measured. Identify the knots on the tracing paper with numerals (1, 2, 3, etc.) that correspond with the outward progression of the prominence.

If the scale of the photos is known, use the millimeter ruler to measure the distance between any two successive knots to find the distance the prominence has ejected. Without knowledge of the print scale of the photographs, but a field of view capturing the center of the disc, the radius of the Sun may be used to calculate the scale. The solar radius is 695,990 km; therefore divide 695,990 by the distance from the disc center to the limb as found with the millimeter ruler to obtain the km/mm scale of the photo.

Velocity (V) is an expression of $V = D/T$, where D is the distance (kilometers) between two points and T is the time (seconds) interval between those points. Consequently, when the formula is solved for the reference points the result is expressed as so many kilometers per second. Outward velocity, away from the Sun, has a positive (+) value; incoming material, which is falling toward the Sun, is given a negative (−) value.

Preparing Photographs and Drawings

Regardless of the wavelength used, solar morphology observations are of two distinct varieties, either wide field or narrow field. A wide field view of ½° or more is identified as "whole disc" and the close-up, narrow field view is "high-resolution". Uniformity of standards is encouraged when it comes to the presentation of photographs and drawings of solar features.

In that way a convenient and valid comparison can be made between observations. With a drawing the information is typically handwritten on the recording form. Photographs, nearly 100 % of which are a digital image, have data embedded in the photo using photo editing software.

Below are a number of guidelines, that if followed will increase the value of any observation you make; some recordable items such as date and time are absolutely essential.

White Light Drawings

- Indicate celestial image orientation, preferably north is upward and the direction of E/W.
- Note the observer's name, address, and date/time of the observation, including the duration.
- Instrumental parameters to include are telescope design and aperture, focal length, eyepiece, as well as filtration. Direct objective or projection screen.
- Seeing conditions, percentage of clouds, wind, and transparency of the sky.
- Carrington rotation.
- High-resolution drawings should include an indication of scale in arc seconds.
- Umbrae are drawn accurately and filled-in with a hard lead pencil. Penumbrae are outlined with a hard lead pencil; in high resolution drawings use a soft lead pencil to shade in penumbral intensities. Faculae are indicated by dashed outlines with brightness noted using the scale: Dim, Bright, or Intense.

Monochromatic Drawings

Although a challenge, it is possible to record the numerous solar events visible to a monochromatic observer. Because there is a considerable amount of detail at even a 5 arc sec resolution, usually more than can be drawn, only the main features should be depicted: filaments, plages, flares, prominences, and sunspot umbrae.

- Indicate celestial image orientation, preferably north is upward and the direction of E/W.
- Note the observer's name, address, and date/time of the observation, including the duration.
- Instrumental parameters to include are telescope design and aperture, focal length, eyepiece, as well as filtration characteristics.
- Seeing conditions, percentage of clouds, wind, and transparency of the sky.
- Carrington rotation.
- High-resolution drawings should include an indication of scale in arc seconds.
- Sunspot umbrae are drawn accurately and filled in with a hard lead pencil. Draw the umbra first, in order to establish reference points for locating the positions of other features accurately.
- After the umbrae are located, concentrate on the other features with the focus being: *position, shape, size, and relative brightness*.
- A filament is drawn using a short or long, narrow or wide black line shaped as it appears to the eye. A plage or flare is drawn with a closed line similar to the white light penumbra.

- Draw the prominence as a black outline, include position angle if possible.
- After all features are rendered on the drawing, the relative darkness and brightness of each is to be noted next to the feature (see below).

Brightness and Darkness Scale

On a drawing next to the monochromatic feature, mark down the corresponding numerical code that represents its brightness or darkness. Plage and prominences use the same brightness scale and flares use the brightness scale of plage/prominences with yet another brighter value.

If a written description of an event is to be communicated, a shorthand system for note taking is recommended when referring to the feature's brightness/darkness. Filament darkness is abbreviated as FD; flare brightness as FB; and prominence/plage brightness as PB. This is a typical example: "bright flare near faint plage with a nearby dark filament." Abbreviation: FB4 flare near PB2 plage with FD4 filament. The logic behind using such a system is that an event progresses rapidly, therefore time is critical and must be saved (Table 4.1).

Whole Disc Photographs

- Indicate celestial image orientation, preferably north is upward and the east to the left.
- Note the observer's name, address, and date/time of the observation.

Table 4.1 Brightness/Darkness Scale for monochromatic drawings

Filaments	Plage/Proms	Flares
1 = Very Faint Gray	1 = Very Faint	1 = Very Faint
2 = Faint Gray	2 = Faint	2 = Faint
3 = Medium Black	3 = Medium Bright	3 = Medium Bright
4 = Deep Black	4 = Bright	4 = Bright
		5 = Very Bright

- Instrumental parameters to include are telescope design and aperture, focal length, as well as filtration characteristics.
- Seeing conditions, percentage of clouds, wind, and transparency of the sky.
- Carrington rotation.

High-Resolution Photographs

- Indicate celestial image orientation, preferably north is upward and the east to the left.
- Note the observer's name, address, and date/time of the observation.
- Instrumental parameters to include are telescope design and aperture, focal length, as well as filtration characteristics.
- Seeing conditions, percentage of clouds, wind, and transparency of the sky.
- Carrington rotation.
- High-resolution photographs should include an indication of scale in arc seconds.

APPENDIX A

Often Asked Questions

Solar observers attract a lot of attention, whether at a club event, astronomical convention, or from neighbors when you are observing in the backyard. This is a natural consequence of the state of affairs: solar observers enjoy their hobby when the majority of the non-viewing public is out and about. Observers of galaxies and nebulae often keep late night hours when few, if any of the community are available.

At astronomy events solar observers are many times busy demonstrating their wares to fellow astronomy enthusiasts, some who are barely mindful that the Sun can be observed safely. It's not unusual for a veteran amateur astronomer to get his first glimpse of a prominence at one of these get-togethers. Then his interest is piqued, a flood of questions emerges and a new solar observer is born.

Some questions posed to the experienced solar observer seem elementary, others are highly technical, but all are legitimate inquiries. Remember when you were just developing an interest in Sun watching? You may have had those same wonderings. Following are ten questions that have been asked of experienced solarphiles, and answers that delighted the inquiring new observer.

What's the Difference Between a Surge and a Spray Prominence?

The surge and spray are both prominence ejection events. A surge will often burst outward at any angle from vertical to horizontal, attaining a

Jamey L. Jenkins, *Observing the Sun: A Pocket Field Guide*,
Astronomer's Pocket Field Guide, DOI 10.1007/978-1-4614-8015-0,

possible elevation of several 100,000 km, lose inertia and then fall back onto itself with a great splashing effect. The velocity of a typical surge is less than 200 km/s, less than the escape velocity of the Sun, which at the photosphere is about 600 km/s.

Two categories of spray are recognized: the *flare spray*, resulting in the expulsion of plasma due to the violent effect of a flare and the *prominence spray*, which is the eruption of an overlying filament above a neutral line. The prominence spray is the less energetic of the two events. The velocities of many sprays exceed 200 km/s and have been known to go beyond 2,000 km/s, easily attaining escape velocity from the Sun's gravity.

The difference between a surge and a spray in a nutshell is this: A surge is *controlled* or *collimated*, a relatively continuous expulsion of material, blown away by an underlying low-level compact, or simple-loop solar flare. A surge is collimated because it is bound by the magnetic field of the AR's primary sunspot. While a surge is a continuous burst of collimated material following the primary spot's magnetic field lines, the spray is a single ejection of filament material in an *uncontrolled* pattern, with matter going in various directions. Again a solar flare is the progenitor, but in this case the flare is much larger and within a less commanding magnetic field. The surge can be compared to a "bottle rocket" and the spray to an exploding "aerial bomb."

Do Sunspots Disappear Completely During Sunspot Minimum? How Many Are Visible at Maximum?

Disappearing sunspots? Sometimes it appears so, but not often. The observing circumstance and instrument employed play a big role into the number of spots visible. For example, compare the results of an Earth bound observer using a 50 mm aperture telescope verses a space based solar observatory telescope. Maybe the Earth observer can see 15 sunspots while the orbiting telescope spots 50. Which data is correct for a sunspot count? The point is that there may be days when the Earth bound observer cannot seem to locate a sunspot simply because it is beyond his ability to do so.

Solar Cycle 23 had a solar minimum that included some of the blankest appearing solar discs in the last fifty years. It was reported that during this cycle over 800 days had a spotless Sun. The years 1645–1715 are called the Maunder Minimum because it was a period when very few sunspots were observed. While activity on the Sun was indeed low, it didn't come to an abrupt end – even during the Maunder Minimum, a number of days had sunspots however few they were.

During solar maximum, sunspot production can be difficult to predict. Solar Cycle 23 while notorious for its low output had a smoothed maximum sunspot number in excess of 120. A typical solar cycle can easily be twice that amount.

Why do You Observe the Sun?

That is a question that has been asked *many* times, and everyone has a different personal response. Some individuals speak of Sun observing as being a humbling and relaxing experience. One observer has quipped humorously that the Sun is easy to find in the sky (true!). Another said it is just fun to share with the public. Fun and fascination of examining a star close up and in intimate detail is a common theme among die-hard solar observers. Another reason the Sun is a favorite is its shear dynamics. There is *always* something happening on the Sun. And many of these events are truly awe-inspiring to witness.

In what other astronomical pursuit can such an abundance of spectacular activity occur in so short a time span? It is for these reasons, and others that many amateur astronomers study the Sun.

How Come Prominences and Flares Are Not Visible in My Telescope?

Depends on your telescope and the accessories it uses, most likely you are referring to a telescope used primarily for sunspot observing. Features such as prominences shine brightest in wavelengths in which they emit light (called emission). While these features can become bright enough to be visible in "white light", they normally fail to achieve anywhere near that intensity level.

The real problem is that the photosphere, where sunspots develop, is so much brighter than the chromosphere where prominences occur, that the chromosphere becomes dominated by the overwhelming brightness of the photosphere. This is a lot like trying to see a burning match placed before the concentrated beam of a high-powered searchlight. The beam from the searchlight dwarfs the tiny flame unmercifully, just as does the photosphere when we attempt to see the much weaker chromosphere. What is needed is a way to effectively *turn off* the photosphere. While there is no light switch to accomplish that, what we can do is to block its excess light, and in that way isolate the chromospheric light.

Modern solar observers accomplish this by viewing the Sun through a special narrowband filter. Excess light originating in the photosphere is filtered out, leaving only the light of the chromosphere to dominate the view. Such a view is called monochromatic, because that view is of only one color.

What's Best for Observing Sunspots, a Projection Screen or an Objective Filter?

If your goal is producing a whole disc drawing of sunspots or determining heliographic coordinates of a white light feature, then a solar projection screen is the way to proceed. Drawing sunspots (tracing or transferring them from a grid) with a projection screen is far more accurate than free-hand drawing a filtered view seen through an eyepiece. A projection device, though deemed the safest observing technique, however has its drawbacks in that ambient light falling on the viewing surface washes out some detail that could otherwise be visible.

Objective filters when attached securely to the telescope will provide a safe and superior view of sunspot detail. Supplementary filters also can be combined with the direct observing package to enhance a view of specific features. If imaging the photosphere interests you then an objective filter will be inescapable.

To briefly summarize, positional studies are best done by projection, morphological studies by direct viewing, sunspot counting can be done either way, and the safety differences between the two techniques are essentially nil.

Why Do H-alpha Filters Have to Cost So Much?

Find solace in the fact that such devices cost far less now than they did in the past. Market forces and competition make products available to many more observers than would have been possible otherwise.

In reference to the question, manufacturing procedures probably contribute more than anything to the cost. A filter has to be very precisely coated and manufactured to extreme tolerances to pass only very narrow wavelengths of light. The filters only operate properly within specific temperature ranges, necessitating the creation of a controlled environment, another costly factor.

Technology necessitates the cost, and when all things are considered the price for your safety, and the view it provides make such an appliance affordable.

When Is the Best Time to Observe the Sun?

"Whenever it is above the horizon!" is the often-quoted answer, but the most wanted response is usually related to seeing conditions and the Sun. The Holy Grail of solar observing is this, "Seeing is everything!"

Astronomical seeing is defined as the quality of the atmosphere between the observer and what is being observed. It is the state of the medium the waves of light must transverse to reach your eyes. Daytime seeing creates situations where the Sun may appear perfectly still and richly detailed or all awash, smeared to a point where only large sunspot umbra is discernible. As you can therefore imagine seeing conditions are very important to solar observing.

A few "urban legends" exist pertaining to daytime seeing. One is that the very best time to observe the Sun is in the early morning, before the surroundings have had an opportunity to heat up and disturb the local air. This can be true, for some locations, but not always. The reality is that superior seeing conditions, day or night, vary from location to location and often hourly within individual locales. Some observers experience superior daytime seeing when the Sun is high in the sky, when light from the Sun has less disruptive atmosphere to pass through. The direction and strength of the wind has an influence on seeing conditions, as does the temperature gradient. The closer together the daytime and nighttime air temperature is, the less turbulent the seeing at times appears. And yes, occasionally early morning solar observing *is* preferred.

To discover how local seeing conditions are affected study your local seeing conditions and weather patterns, then note when prime solar observing occurs for you. Is the air steadier after a cold or warm front has passed? Is there a part of the day when the Sun appears less turbulent? There is no rule of thumb for the time of day that is always best for solar observing. What is true is that daytime seeing varies with the conditions present at any given site. Study those conditions during superior seeing, and when they repeat, take advantage of the opportunity.

Are Prominences Really Reddish-pink at a Total Solar Eclipse?

Yes, they truly are. In fact prominences appear that color because they are in emission and glow brightest at 656.3 nm in the solar spectrum, the location of the reddish-pink H-alpha line. The color of the chromosphere is owed to the dominance of the H-alpha line and its location in the red part of the spectrum.

Prominences also emit light at other wavelengths, such as in Calcium in the blue part of the spectrum at 393.3 nm. This emission is much weaker and contributes feebly to a prom's visual color at an eclipse.

Are You Sure that East on the Sun Is Toward the Left, Geographic Maps Always Have West on the Left and East on the Right?

Directions on the Sun are defined by the celestial directions of N-S-E-W, which are extensions of the Earth's equator and axis of rotation. Photos are usually created with east to the left, roughly matching the northern hemisphere's appearance of the Sun on the meridian.

The simplest way to determine celestial directions is to center the Sun in the telescope's field of view and nudge the telescope toward the north celestial pole; the Sun's northern limb will begin to move toward the center of the field. East and west are determined by allowing the Sun to drift across the field of view. Turn off the telescope's driving mechanism and watch the Sun drift slowly out of the eyepiece. The first limb to exit is the preceding limb, this locates west, and the east limb follows, leaving the view last.

How Can I Get into Observing the Sun?

Do some research and understand what solar safety is about. *Never attempt to view the Sun through any optical instrument that has not been properly fitted with safe solar observing appliances.*

Start off with white light observing, the photosphere offers numerous opportunities for discovering how the Sun works, and helps you develop the skills necessary for increasingly difficult observations. The least expensive and perhaps the safest method of observing the Sun in white light is the projection method. A refractor or Newtonian reflecting telescope is suitable for solar projection. Sunspot counting is how many solar observers begin; give it a try, and then graduate to drawing or imaging active regions. After having learned the basics, move on to monochromatic observing in Ca-K or H-alpha and explore the solar chromosphere.

Locate and establish contact with other solar observers through organizations like the AAVSO, ALPO, or BAA. Locally, many clubs these days have at least a few members who enjoy solar observing, seek them out for advice and guidance. This also gives you the opportunity to evaluate equipment and accessories before making a purchase of your own.

The Sun provides an alternative to sleepless nights, cold numb fingers, and straining your eyes to glimpse an inert faint fuzzy. The fact remains that all events on the Sun are unique, never to be exactly repeated, and because of this your observations have a special value. So, investigate the Sun, give it the respect it's owed, and enjoy one of nature's most magnificent spectacles from your own backyard.

APPENDIX B

Amateur Observing Organizations

Association of Lunar and Planetary Observers Solar Section

www.alpo-astronomy.org/solarblog/

American Association of Variable Star Observers Solar Section

www.aavso.org/solar

British Astronomical Association Solar Section

britastro.org/baa

Belgian Solar Observer

www.bso.vvs.be/index_en.php

Solar Database

basehalpha.ift.fr/

Useful Personal Solar Websites

Peter Meadows provides a websites with information pertaining to daily sunspot drawings and freeware solar software, including Stonyhurst Discs.

www.petermeadows.com/indexsolar.html

Jamey L. Jenkins, *Observing the Sun: A Pocket Field Guide*,
Astronomer's Pocket Field Guide, DOI 10.1007/978-1-4614-8015-0,

Art Whipple produces a site featuring superb sunspot imaging and information on his techniques.

home.comcast.net/~jim6/sunspots.htm

Jim Ferreira has a website containing spectacular videos and picture-series that illustrate the dynamics of H-alpha observing.

www.lafterhall.com/astro.html

Christian Viladrich images the Sun in Ca-K and white light and posts top notch samples on his web pages.

christian.viladrich.perso.neuf.fr/index.html

Harry Roberts has an informative blog page that contains observational drawings made by Harry of the latest solar events.

www.harryrobertsastronomy.com/

Spaceweather details daily solar activity and provides a daily image with AR numbers superimposed.

www.spaceweather.com

SOHO is the Solar and Heliospheric Observatory, a spacecraft designed to study the Sun from its deep core to the outer corona and the solar wind. Up to date images are provided in various wavelengths.

sohowww.nascom.nasa.gov/

Mees Solar Observatory supplies daily active region maps.

www.solar.ifa.hawaii.edu/ARMaps/index.shtml

Mt. Wilson Solar Observatory publishes a daily sunspot drawing from the 150-ft. tower telescope.

obs.astro.ucla.edu/cur_drw.html

APPENDIX C

Glossary of Solar Terms

Absolute zero The coldest temperature, zero degrees Kelvin.

Absorption lines The dark lines crossing a spectrum caused by the absorbing of photons as electrons jump to a higher energy level.

Active Region Filament (ARF) A filament/prominence in close proximity to an active region, often magnetically associated with that region.

Active region Location in the photosphere in which the formation of sunspots, faculae, etc. arise over time.

Angstrom Unit of measure for the expression of the wavelength of light equal to 0.1 nm or one ten-millionth of a millimeter.

Annular eclipse An eclipse of the Sun that at maximum phase produces a ring of sunlight.

Aperture The diameter of the objective of a telescope.

Arc second Angular measurement whereas 60 arc sec equal 1 arc min. Sixty arc minutes are equal to 1 arc degree or roughly twice the diameter of the Sun or Moon.

Arcade A series of magnetic loops found on the Sun.

Aurora Glowing gases in the Earth's upper atmosphere, excited by solar particles originating in the Sun and carried to Earth by the solar wind.

Baily's beads An effect occurring just before and after a total eclipse of the Sun, as sunlight passes between mountain peaks found on the lunar limb.

Balmer series A pattern of spectral lines in the visible spectrum produced by the jumping of electrons from one energy level to another in hydrogen.

Bandpass Measurement between the lower and upper cut off frequency of an optical filter, usually measured at the full-width half-maximum point.

Jamey L. Jenkins, *Observing the Sun: A Pocket Field Guide*,
Astronomer's Pocket Field Guide, DOI 10.1007/978-1-4614-8015-0,

Bandwidth See bandpass.

Bipolar sunspot Two concentrations of umbral spots or sunspots having a positive and negative magnetic polarity with a minimum separation of three heliographic degrees.

Bo A parameter for the calculation of heliographic coordinates representing the nod or changing latitude at the center of the solar disc.

Bow shock The Sun facing edge of the Earth's magnetosphere.

Bright points Miniscule features that result from intensive, but small flux tubes (300 km) poking through the photosphere allowing an inspection of the solar interior. A group of bright points, forming a string or chain upon the intergranular wall, are known as filigree, observable in the solar continuum, but more frequently at the G-band (430.5 nm)

Broadband Term for a filter transmitting a wide bandwidth (i.e. 100 Å or more).

Buoyancy The upward force that arises from an increase in fluid force at a greater depth.

Bush A cluster of spicules seen in absorption on the solar disc.

Butterfly diagram The graphical representation depicting the latitude of emerging sunspots verses the time progression of a solar cycle.

Calcium-H and K The spectral lines located at 396.8 nm (H) and 393.3 nm (K), also known collectively as Ca-II.

Carrington Rotation The number of rotations of the Sun as seen from Earth since 9 November 1853.

Celestial directions Cardinal directions N, S, E, and W projected on the celestial sphere that coincide with the Earth's equator and axis of rotation.

Center Wavelength (CWL) The wavelength found at the midpoint of the full-width half-maximum.

Central Meridian (CM) An imaginary line drawn from the north to the south pole of the Sun.

Chain A row or line of spicules seen in absorption on the solar disc.

Chromosphere The layer of solar atmosphere directly above the photosphere and below the corona.

Chromospheric network A web-like mesh covering almost the entire Sun and displaying a bright pattern in Ca-K and in H-alpha, a dark one.

Coelostat Two mirror system reflecting a stationary image of the Sun to a telescope. Also see heliostat.

Compact flare A flare which is closely packed yet intense.

Continuum The areas of the spectrum that are neither in emission or absorption.

Convection zone An inner layer of the Sun in which energy transfer occurs through convection.

Core (Sun) The central region of the solar interior powered by the hydrogen to helium fusion process.

Core (sunspot) The umbral area of a sunspot umbra having the greatest magnetic strength.

Corona The outer atmosphere of the Sun, beyond the chromosphere.

Coronagraph An instrument designed to permit viewing of the Sun's corona.

Coronal Mass Ejection (CME) A large eruption of particles from the Sun.

Coronal streamer A loop shaped feature of the lower corona.

Diamond ring During a total eclipse, an effect caused by a solitary shimmer of sunlight passing between mountain peaks at the lunar limb a single Baily's bead.

Differential rotation The lack of singularity in a rotation period due to the liquid-like nature of the body.

Diffraction grating A finely grooved substrate whose purpose is the dispersion of light.

Disparition brusque The sudden disappearance of a filament (prominence).

Dobsonian Solar Telescope A unique Newtonian style telescope designed for white light solar observing, the creation of sidewalk astronomer, John Dobson.

Doppler shift The stretching or compressing of spectral lines due to the approaching or receding of an object.

Double stack A method of narrowing the bandwidth of an etalon by the addition of a second etalon.

Ellerman bomb A small bright feature visible in the wings of H-alpha notably around a sunspot. Circular with a diameter less than 3 arc sec, they have a lifetime of a few minutes to several hours. They are known in some circles as moustaches.

Emerging Flux Region (EFR) An area of new magnetic flux on the Sun.

Emission lines The bright lines crossing a spectrum caused by the emitting of photons as electrons jump to a lower energy level.

End-loading A monochromatic narrowband filter which attaches at the exit of a telescope.

Energy Rejection Filter (ERF) A pre-filter that is placed over the opening of a telescope for the purpose of absorbing or reflecting UV/IR light and reducing the heat load on the interference filter.

Ephemeral region Tiny short-lived active region.

Ephemeris Tables that are published yearly listing the daily orientation of the Sun for the factors *P*, *Bo*, and *Lo*.

Eruptive prominence An active prominence being ejected from the Sun.

Etalon An optical filter that operates by the multiple-beam interference of light, reflected and transmitted by a pair of parallel flat reflecting plates.

Evershed effect Penumbral gas flow pattern.

Facula A luminous, cloud-like patch or venous streak of material seen in white light surrounding or near a sunspot.

Fibril Tiny dark appearing structure which follows a magnetic field line sometimes attached to a prominence seen on the solar disc. Analogous to a mottle longer than a few seconds of arc.

Field angle The angle of outside light rays entering a telescope. One example is illustrated by the angular size of the Sun as it appears in the sky.

Filament channel An opening of fibrils underlying a filament.

Filament In monochromatic light a prominence viewed on the disc of the Sun. In white light a structure that radiates about an umbra like fine dark threads.

Filar micrometer A tool for measuring angular displacement through an optical instrument.

Filigree Tiny bright flux tubes popped through the solar surface with a diameter of about 150 km.

Flash phase The period of rapid H-alpha brightening experienced in a solar flare.

Flux tube A strand or kink of magnetic field suspended in the convection zone.

Foot point The location where a magnetic loop meets the photosphere.

Fraunhofer line An atomic line visible in a spectrum.

Front-loading A monochromatic narrowband filter which attaches at the entrance of a telescope.

Full-Width Half-Maximum (FWHM) The measured width of the bandpass, in nanometers or angstroms, at one-half of the maximum transmission.

G-Band The location of several spectral lines that go into emission during a flare at about 430 nm.

Granulation The textured pattern found over the entire photospheric surface. See granule.

Granule The top of a rising column of gas, originating deep within the convection zone of the Sun.

H and K lines See Calcium H and K.

Halo The glowing effect that surrounds the Sun experienced when a CME is directed toward the Earth.

H-alpha The spectral line located at 656.3 nm.

Hedgerow prominence Quiescent form of prominence generally resembling a row of trees or haystacks.

Heliographic coordinates The system of latitude and longitude on the solar disc.

Helioseismology The study of low frequency sound waves originating in the Sun.

Heliostat A single or multiple mirror system reflecting an image of the Sun to an optical instrument.

Helmet streamer A pointed streamer formation found in the corona.

Helmholtz contraction The process of turning gravity's energy to heat as induced by density and pressure.

Herschel wedge A narrow prism used for safely observing the white light Sun when combined with suitable supplementary filters.

Hossfield pyramid The name given a pyramid shaped projection box for white light solar observing devised by Casper Hossfield of the AAVSO Solar Division.

Hybrid eclipse An eclipse of the Sun which shares characteristics of both an annular eclipse and a total solar eclipse.

Hyder flare A flare that occurs some distance from an active region, in the quiet Sun, and has a relationship to the sudden disappearance (disparition brusque) of a filament.

Hydrogen The most plentiful element in the Sun and universe, made of one proton and one electron.

Hydrostatic equilibrium The balance of outward pressure from the compressed gas of the Sun and the solar gravity without which the external layers of the Sun would collapse onto the core.

Infrared (IR) Electromagnetic radiation with a wavelength between approximately 700 and 1,400 nm.

Inner bright ring A brightening within the penumbra located between the umbra and penumbra at a rough region where penumbral filaments have the appearance of extensions of the umbra.

Instrument angle The angle of light rays converging to a focus in a telescope.

Interactive prominence A type of prominence exhibiting material flow that is nearly horizontal between individual components.

Interference filter An optical appliance with several layers of evaporated coatings on a substrate, whose spectral transmission characteristics are the result of the interference of light rather than absorption.

Intergranular wall The intergranular wall is what defines the shape of a granule.

Ion An atom having one or more electrons missing.

Ionization The stripping of electrons from an atom.

Irregular penumbra A sunspot penumbra which has been mutated by complex magnetic fields.

Isophote contour map A graphically created image that interprets the many levels of density within a photo.

Kelvin A unit of temperature in which zero Kelvin is based on −273.15 °C or absolute zero.

K-grains Observed in monochromatic Ca-K or Ca-H light and have a correlation to and similar appearance to bright points.

Kirchhoff's Laws Of spectroscopy state that (1) A hot dense glowing body, produces a continuous spectrum lacking spectral lines. (2) View a continuous spectrum through a cooler, transparent gas and dark lines called absorption lines appear. (3) Hot transparent gas before a cooler background emits the bright spectral lines we call, emission lines.

Light bridge Any material brighter than an umbra that also divides the umbra, often times dividing even a penumbra.

Limb darkening The decrease in intensity of the Sun as one approaches the solar limb.

Limb flare The appearance of a solar flare and its associated events viewed in profile at the solar limb.

Lo A parameter for the calculation of heliographic coordinates representing the longitude of the Sun's central meridian. *Lo* is 0° at the beginning of each new solar rotation.

Lyot filter A monochromatic filter that produces a narrowband transmittance via the principle of birefringence.

Magnetic cycle The return of the magnetic polarity of sunspots within a given hemisphere to the polarity experienced prior to a new solar cycle, a period of approximately 22 years.

Magnetic field An area exhibiting a magnetic influence.

Magnetic reconnection The process whereby magnetic field lines break and reconnect with other field lines. Also known as re-alignment.

Magnetopause The inner borderline nearest the Earth of the magnetosheath.

Magnetosheath Location between the bow shock and the magnetopause here the bulk of the solar wind is repelled around the Earth's magnetic field

Magnetosphere The vicinity around the Earth dominated by its magnetic field.

Magnitude A measure of brightness of a celestial body.

Major flare A large flare event that involves the development of two or more ribbons of H-alpha emission.

McIntosh classification A three-digit white light sunspot classification scheme devised for flare prediction.

Mean Daily Frequency (MDF) An index of solar activity determined by the number of Active Regions visible as the average for a monthly period.

Menzel-Evans-Jones classification A prominence classification system based whether a prominence is ascending or descending in the chromosphere, its relationship to any nearby sunspot, and the general appearance of the prominence.

Meridional flow A gradual looping movement of plasma in the convention zone from the equatorial to the polar region and back again.

Monochromatic One-color, as when referring to light from the H-alpha or Calcium K-line.

Moreton wave Shock wave visible in the chromosphere radiating from a large flare.

Morphology The study of the changing appearance of the Sun.

Mottle A spicule seen against the solar disc. Also see fibril.

Nanometer The nanometer (nm) is a unit measurement of wavelengths of electromagnetic radiation (light). One nanometer is equal to one billionth of a meter (1×10^{-9} m).

Narrowband Term for a filter transmitting a narrow bandwidth (i.e. less than 100 Å).

Neutral line The area where an Active Region's magnetic field reverses polarity.

Normal incidence Light rays that have a normal or parallel path.

Objective filter White light solar filter that mounts at the entrance to a telescope.

Occulting cone Polished metal cone shaped devise to block the light from the solar disc in a prominence telescope.

Off-axis In solar observing the placement of a sub-diameter objective filter at the side of the optical axis of a telescope to avoid the internal optical components (secondary mirror, mounting hardware, etc.) from being within the incoming light path.

Outer bright ring The brightening and aligning of the granules encircling the outer edge of a sunspot, beyond the penumbra.

Oven An electrically controlled heating device to regulate the operating temperature of a narrowband monochromatic filter.

P Parameter for the calculation of heliographic coordinates representing the displacement the north rotational axis of the Sun relative to the rotational axis of the Earth.

Peak transmission The maximum percentage of transmission found within the bandwidth.

Penumbra The lighter, grayish outer region surrounding the umbra in a sunspot.

Penumbral filaments Structures of fine dark threads that radiate about a sunspot umbra.

Penumbral grains Bright regions located between penumbral filaments.

Photon A particle that supports electromagnetic radiation.

Photosphere The lowest layer of the Sun's atmosphere. The region contains sunspots, granulation, and faculae.

Plage A lower chromosphere feature in the habit of surrounding a sunspot as a bright cloud-like form. It marks the location of the magnetic field associated with the sunspot.

Plasma A brew of ions and electrons that react energetically with magnetic fields.

Polar crown The appearance of several filaments linking together in an east–west direction to form one long strand of filamentary material in a high solar latitude.

Polar faculae Small bits of faculae forming in high latitude regions outside the sunspot zones.

Polar plume A lesser height coronal streamer located near the north and south poles. Unlike the closed loops of the helmet streamer these structures result from open magnetic field lines.

Pore A tiny structure, with a diameter from 1 to 5 arc sec, darker than a granule but brighter than the umbra of a well-developed sunspot.

Position Angle (PA) The angular offset in degrees around the limb of the Sun defining the position of a prominence or other feature. The cardinal directions are N = 0°, E = 90°, S = 180, and W = 270°.

Post flare loop The loop system seen typically following a two-ribbon flare.

Prominence A cloud of gas suspended above the surface of the Sun.

Proton-proton cycle The process experienced by a star equal to or less than the mass of the Sun by which it converts hydrogen to helium.

Protosun A region of greater density and pressure near the center of the solar nebula.

Puff Also called a *smokestack*, the puff is a smaller variation of the classic surge, with an appearance suggestive of its name. Visible as a blue shifted feature with a lifetime of several minutes, they are seen in an active region as well as in an emerging flux region.

Quiescent prominence A quiet behaving prominence that changes its appearance only moderately with time.

Quiet Region Filament (QRF) A filament appearing in the quiet of the Sun, separate from an active region.

Radiative zone An inner layer of the Sun in which energy transfer occurs through radiative properties.

Relative sunspot number An index of solar activity determined by the number of sunspot groups and individual sunspots visible on the face of the Sun.

Reversing layer A layer of cooler gas near the lower chromosphere which gives rise to absorption lines in the solar spectrum.

Rosette A pattern of spicules seen in absorption on the solar disc resembling a flower with petals.

Rudimentary penumbra The beginning phase of a penumbra, forming from the intergranular material, surrounding a newly developed umbra.

Seeing conditions The quality of the atmosphere between an observer and what is being viewed.

Shadow bands Very low contrast, light and dark wavy forms that appear just prior to and following the totality phase of a solar eclipse. It is the manifestation of seeing effects within the lower atmosphere.

Shear To cause to move along the plane of contact

Sidereal period The time required for the Sun to complete one rotation, as seen from a fixed point in space.

Solar continuum Basically a view of the Sun encompassing a wide bandwidth of visible light. It would be analogous to a white light view. Also see continuum.

Solar cycle The approximate 11-year rise and fall in solar activity.

Solar flare The swift release of energy that has accumulated within the magnetic field of an Active Region.

Solar maximum The peak of solar activity during the 11-year solar cycle.

Solar minimum The time that experiences little activity during the 11-year solar cycle.

Solar nebula A vast cloud of gas and dust from which it is believed the Sun and solar system originated.

Solar projection A technique for viewing the white light Sun by projecting an enlarged image of the solar disc onto a white screen some distance from the eyepiece of a telescope.

Solar wind The constant stream of particles flowing from the Sun into outer space.

Spaceweather The status of the space environment near the Earth, as it has been affected by the release of energy and particles from the Sun.

Spectroheliosocope An instrument that synthesizes a monochromatic view of the Sun.

Spectroscope An instrument for the dispersion of light.

Spectrum The result when electromagnetic energy is dispersed into its constituent rays by order of wavelength, for example, a rainbow.

Spicule A fine structure resembling a tiny gas jet at the solar limb. On the disc in H-alpha it appears dark and then is known as a mottle or fibril.

Spray The dramatic and uncontrolled eruption of a prominence into a sporadic-like form.

Star A giant sphere shaped ball of gas that through nuclear reactions releases energy in its core.

Stonyhurst disc A template or grid that shows the lines of heliographic latitude and longitude relative to a given value of *Bo*.

Subflare A compact flare having an area equal to or less than 2.0 heliographic square degrees.

Sunspot A dark region on the photosphere that results from convection being stifled by magnetic fields within the region.

Sunspot drift The passage of a sunspot or sunspot group from east to west as seen in a stationary telescope, caused by the Earth's rotation.

Sunspot group A clump of sunspots.

Sunspot number An index for measuring solar activity, $R = 10\,g + s$. Also known as the Wolf number from its originator, Rudolph Wolf.

Sunspot zone A region of approximately 35° on both sides of the solar equator in which sunspots form.

Supergranulation The large scale pattern in the photosphere of organized cellular structure. Each cell contains hundreds of individual granules and has a diameter around 30,000 km. The chromospheric network overlays the supergranulation pattern.

Superpenumbra An effect experienced with large sunspots caused by fibrils diverging around a spot, giving the appearance of the thickening of the penumbral region.

Surge The rapid eruption of a prominence following and constrained by a magnetic field into a jet-like form.

Synodic period The apparent rotation of the Sun, as seen from the orbiting Earth.

Tachocline The shearing region located between the radiative and convection zones of the solar interior.

Telecentric lens A supplementary lens system intended to create normal incidence light rays from the converging light rays of a telescope.

Ten-degree rule A general guideline that states: "any spot or cluster of spots that is at least 10° of heliographic latitude or longitude away from any other spot counts as one sunspot group."

Thermal equilibrium The uniform conversion of hydrogen to helium in the Sun.

Tongues The formation of a bipolar sunspot resembling the yin-yang pictogram. The cause is the rotation of the sunspot at a speed exceeding one degree an hour, wrenching the neutral line to a more east–west direction.

Tornado prominence A prominence form having a column or pillar with a stream of matter flowing from its peak. Sometimes this stream is bent over and nearly reaches back to the chromosphere. Another form has a skeletal-type columnar structure, constructed of single, helically bound streamers with no matter emerging from the top. Tornado prominences exhibit rotation.

Total eclipse A condition in which the Moon completely blocks the solar disc permitting viewing of the chromosphere and corona.

Transition zone The region between the chromosphere and the corona, where temperature makes an abrupt leap from 20,000 to 1,000,000 K.

Transparency A characteristic of sky conditions that describes the opacity of the atmosphere as influenced by water vapor, dust, smoke and other atmospheric particles.

Ultraviolet (UV) Electromagnetic radiation with a wavelength between approximately 280 and 380 nm.

Umbra dot A small dot within an umbra having a brightness between that of the umbra and nearby bright points.

Umbra spot A pore that has become larger and darker than other pores, as dark as the typical sunspot umbra.

Umbra The dark cooler region of a sunspot.

Unipolar sunspot A single concentration of umbral spots or sunspots within a three degree or less area.

Universal time Analogous to Greenwich Mean Time, a 24-h timescale.

Visible light Electromagnetic radiation with a wavelength between approximately 400 and 700 nm.

Void A circumstance where granulation is found to be missing, not to be confused with a pore.

Vortex A condition created by exceptionally large sunspot groups as the surrounding fibrils establish large radially swirled regions.

Wavefront error A distortion or aberration to the incoming rays of light caused by factors including atmospheric seeing or poorly made optics.

Wavelength The distance separating two consecutive wave peaks in a beam of electromagnetic radiation (light).

White Light Flare (WLF) A solar flare of such intensity that its light outshines the solar continuum and becomes visible without a monochromatic filter.

White light The combined result of all wavelengths of visible light.

Wien's Law The principle that the wavelength of the dominant color of a blackbody (star), multiplied by its temperature must equal a specific numerical factor.

Wilson effect The apparent concavity of a symmetrically shaped sunspot as it is near the solar limb.

Winking filament The lifting of a filament so that it oscillates several times, actually causing the filament to become Doppler shifted in and out of visibility, the result of a passing Moreton wave.

Wolf number See sunspot number.

X-ray Electromagnetic radiation with a wavelength between 10 and .01 nm.

APPENDIX D

Master Templates

The following pages contain templates that may be photo copied and enlarged if required for your personal use. The first two originals are of the front and backsides of the proposed sunspot counting card system outlined in Chap. 4 and illustrated in Fig. 4.1. Next is a 360° protractor that is used to find the position angle of prominences around the solar limb. The last eight templates are Stonyhurst Discs provided in 1° increment, positive values of *Bo* at one end and negative at the opposite.

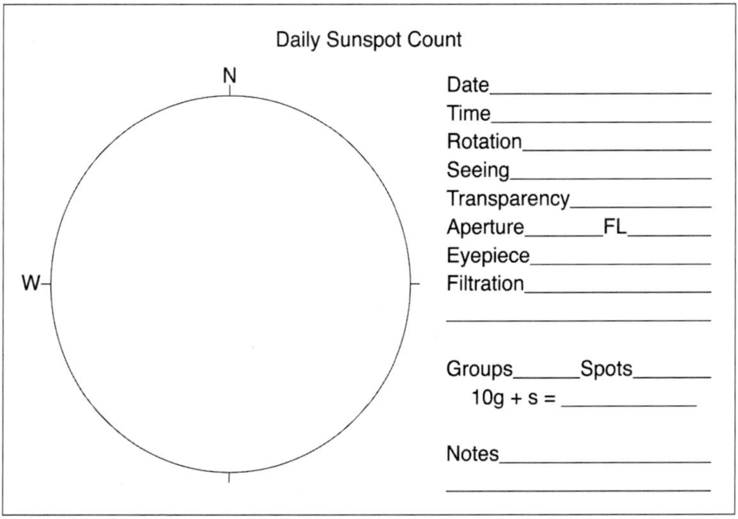

Jamey L. Jenkins, *Observing the Sun: A Pocket Field Guide*,
Astronomer's Pocket Field Guide, DOI 10.1007/978-1-4614-8015-0,

McIntosh Class

A—Individual spot, Unipolar group, No penumbra
B—Bipolar group, No penumbra
C—Bipolar group, One spot with penumbra
D—Bipolar group, Penumbra about both spots, Length less than 10°
E—Bipolar group, Penumbra about both spots, Length 10° to 15°
F—Bipolar group, Penumbra about both spots, Length greater than 15°
H—Individual spot, Unipolar group, With penumbra

Seeing Descriptions

< 1 arc second	Granulation resolved
1-2 arc second	Granulation blotchy
3-5 arc second	Granulation/pores sometimes visible
> 5 arc second	Granulation/pores not visible, ill-defined sunspots

Transparency

Excellent, Good, Fair, or Poor

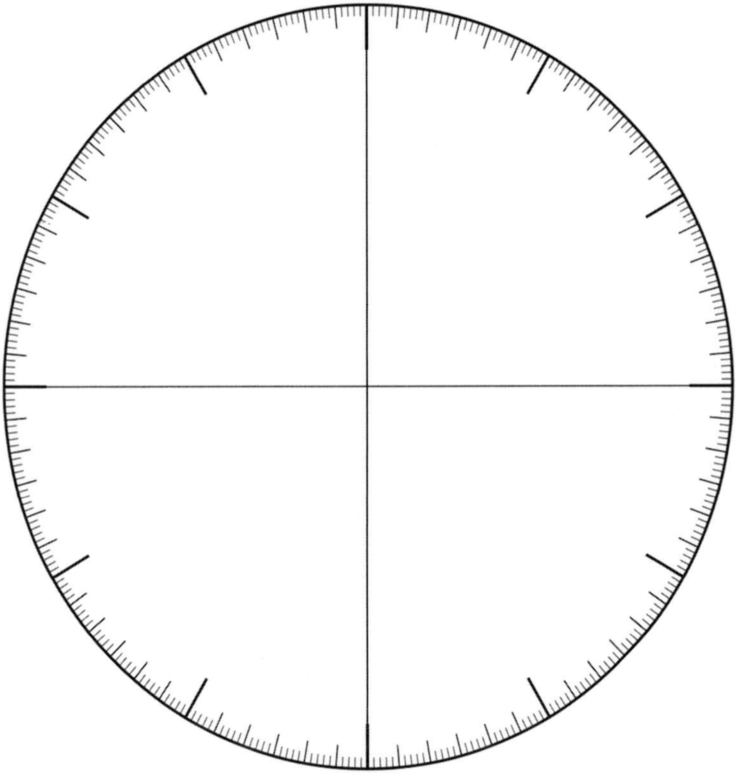

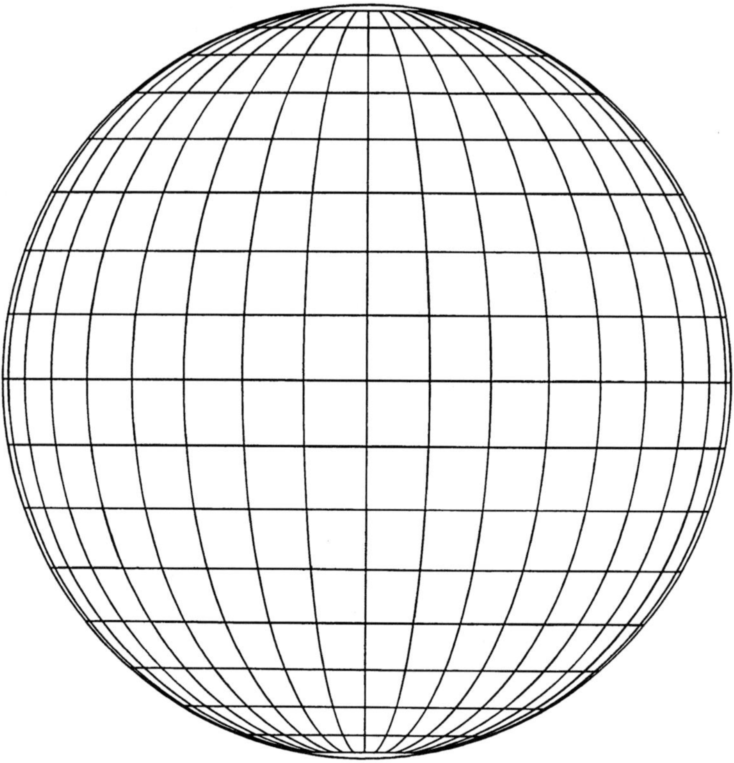

0

-1

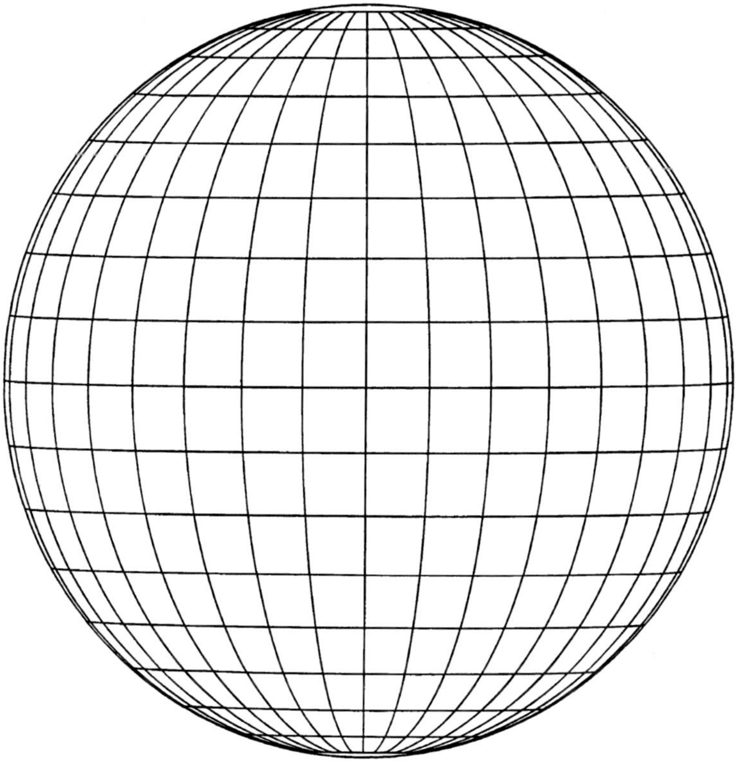

1

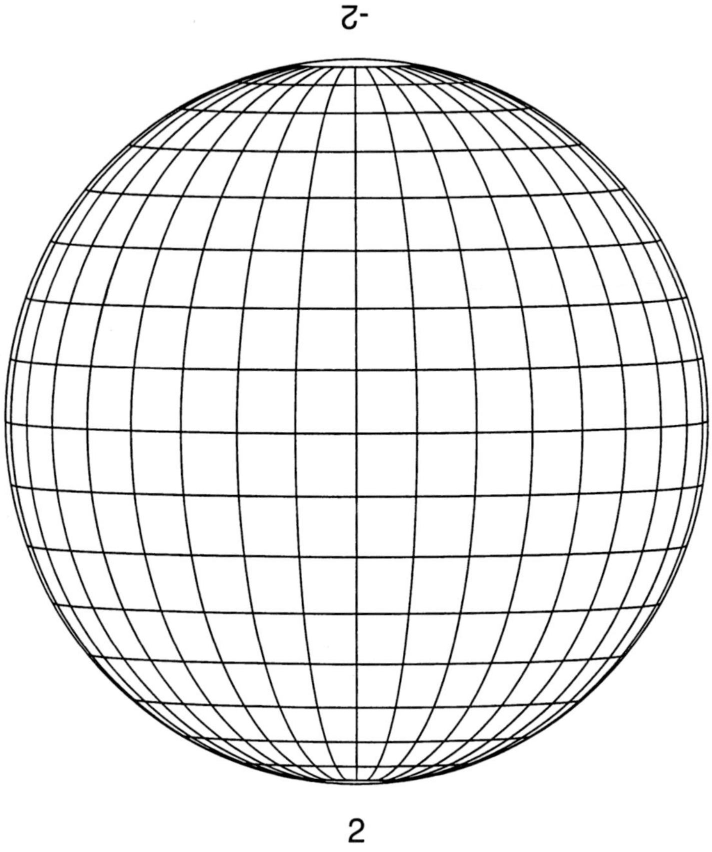
-2
2

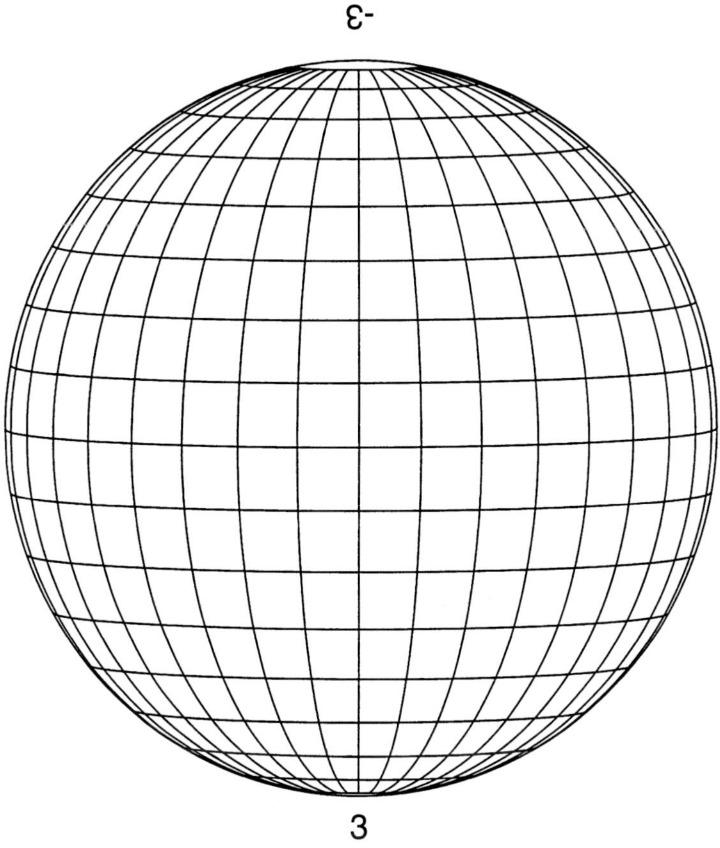
-3
3

-4

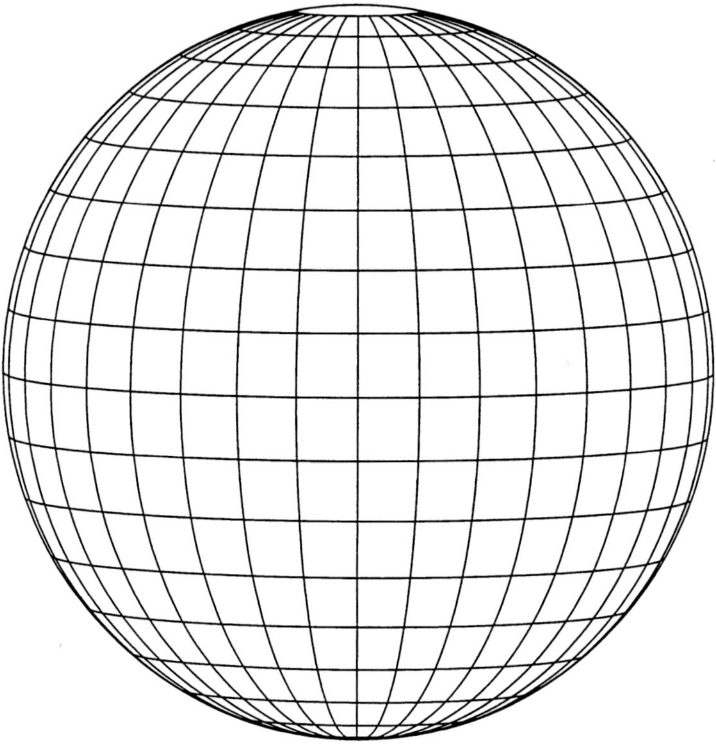

4

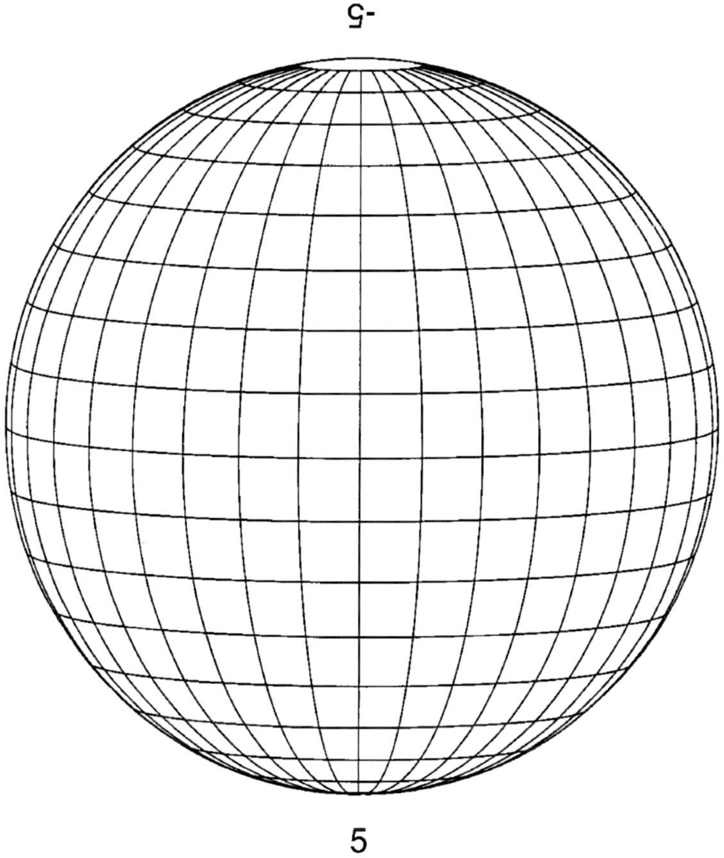
-5
5

9-

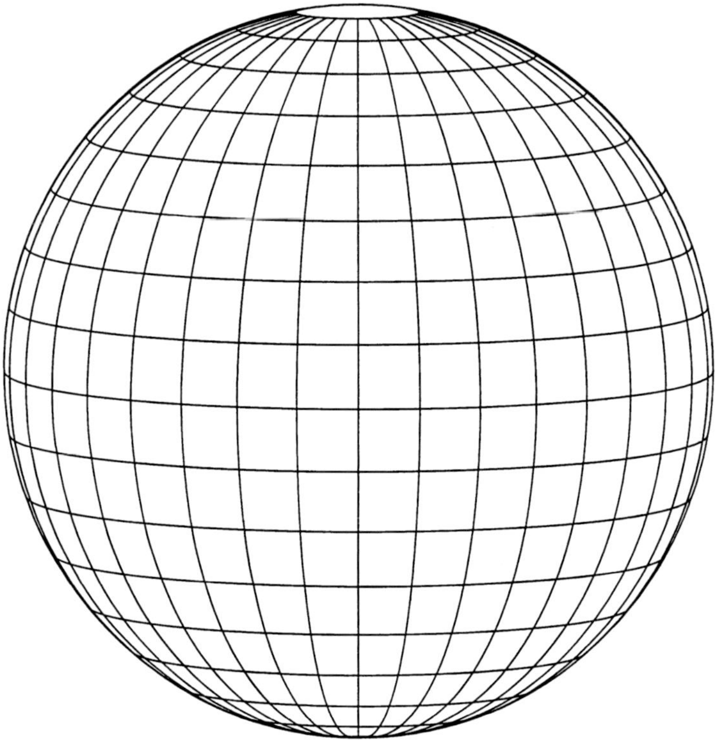

6

-7

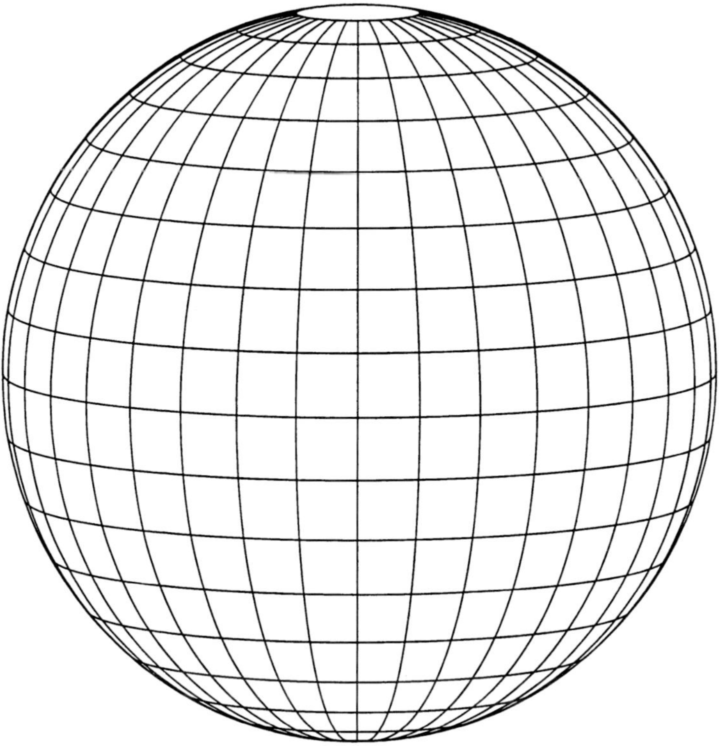

7

APPENDIX E

Daily Solar Ephemeris, June 2013 – December 2016

by Brad Timerson

Association of Lunarand Planetary Observers Solar Section

Calendar	Rotation	Heliographic			Diameter	RA	Dec
Date	Number	*Lo*	*Bo*	*Po*	(arcmin)	(HH:MM)	(deg)
6/1/2013	2137	135.34	−0.66	−15.38	31.543	04:36.8	22.05
6/2/2013	2137	122.10	−0.54	−15.00	31.539	04:40.9	22.18
6/3/2013	2137	108.87	−0.42	−14.62	31.534	04:45.0	22.31
6/4/2013	2137	95.63	−0.30	−14.24	31.530	04:49.2	22.43
6/5/2013	2137	82.39	−0.18	−13.85	31.526	04:53.3	22.54
6/6/2013	2137	69.15	−0.06	−13.46	31.521	04:57.4	22.65
6/7/2013	2137	55.91	0.06	−13.06	31.517	05:1.5	22.75
6/8/2013	2137	42.67	0.18	−12.65	31.513	05:5.6	22.84
6/9/2013	2137	29.44	0.30	−12.25	31.510	05:9.8	22.93
6/10/2013	2137	16.20	0.42	−11.84	31.506	05:13.9	23.01
6/11/2013	2137	2.96	0.54	−11.43	31.503	05:18.1	23.08
6/12/2013	2138	349.72	0.66	−11.01	31.499	05:22.2	23.15
6/13/2013	2138	336.48	0.78	−10.59	31.496	05:26.3	23.21
6/14/2013	2138	323.24	0.90	−10.17	31.493	05:30.5	23.26
6/15/2013	2138	310.00	1.02	−9.74	31.490	05:34.6	23.31
6/16/2013	2138	296.76	1.14	−9.31	31.487	05:38.8	23.35
6/17/2013	2138	283.52	1.26	−8.88	31.485	05:43.0	23.38
6/18/2013	2138	270.28	1.38	−8.44	31.482	05:47.1	23.41
6/19/2013	2138	257.10	1.50	−8.01	31.480	05:51.3	23.43
6/20/2013	2138	243.86	1.61	−7.57	31.478	05:55.4	23.44
6/21/2013	2138	230.62	1.73	−7.13	31.475	05:59.6	23.44
6/22/2013	2138	217.37	1.85	−6.68	31.474	06:3.7	23.44
6/23/2013	2138	204.13	1.96	−6.24	31.472	06:7.9	23.43

(continued)

Jamey L. Jenkins, *Observing the Sun: A Pocket Field Guide*,
Astronomer's Pocket Field Guide, DOI 10.1007/978-1-4614-8015-0,

(continued)

Calendar	Rotation	Heliographic			Diameter	RA	Dec
Date	Number	*Lo*	*Bo*	*Po*	(arcmin)	(HH:MM)	(deg)
6/24/2013	2138	190.89	2.08	−5.79	31.470	06:12.1	23.42
6/25/2013	2138	177.65	2.19	−5.35	31.469	06:16.2	23.39
6/26/2013	2138	164.41	2.31	−4.90	31.467	06:20.4	23.36
6/27/2013	2138	151.17	2.42	−4.45	31.466	06:24.5	23.33
6/28/2013	2138	137.93	2.54	−4.00	31.465	06:28.7	23.28
6/29/2013	2138	124.69	2.65	−3.54	31.464	06:32.8	23.23
6/30/2013	2138	111.45	2.76	−3.09	31.463	06:36.9	23.18
7/1/2013	2138	98.21	2.87	−2.64	31.463	06:41.1	23.11
7/2/2013	2138	84.97	2.98	−2.18	31.462	06:45.2	23.04
7/3/2013	2138	71.73	3.09	−1.73	31.462	06:49.3	22.97
7/4/2013	2138	58.49	3.20	−1.28	31.462	06:53.5	22.88
7/5/2013	2138	45.25	3.31	−0.82	31.462	06:57.6	22.79
7/6/2013	2138	32.01	3.41	−0.37	31.462	07:1.7	22.70
7/7/2013	2138	18.77	3.52	0.08	31.462	07:5.8	22.59
7/8/2013	2138	5.54	3.62	0.53	31.463	07:9.9	22.48
7/9/2013	2139	352.36	3.73	0.98	31.463	07:14.0	22.37
7/10/2013	2139	339.12	3.83	1.43	31.464	07:18.1	22.24
7/11/2013	2139	325.88	3.93	1.88	31.465	07:22.1	22.11
7/12/2013	2139	312.65	4.03	2.33	31.466	07:26.2	21.98
7/13/2013	2139	299.41	4.13	2.78	31.467	07:30.3	21.84
7/14/2013	2139	286.17	4.23	3.22	31.468	07:34.3	21.69
7/15/2013	2139	272.93	4.33	3.67	31.470	07:38.4	21.54
7/16/2013	2139	259.70	4.42	4.11	31.471	07:42.4	21.38
7/17/2013	2139	246.46	4.52	4.55	31.473	07:46.4	21.21
7/18/2013	2139	233.23	4.61	4.99	31.475	07:50.5	21.04
7/19/2013	2139	219.99	4.70	5.42	31.477	07:54.5	20.86
7/20/2013	2139	206.76	4.80	5.86	31.479	07:58.5	20.67
7/21/2013	2139	193.52	4.89	6.29	31.482	08:2.5	20.48
7/22/2013	2139	180.29	4.97	6.72	31.484	08:6.5	20.29
7/23/2013	2139	167.05	5.06	7.14	31.487	08:10.4	20.09
7/24/2013	2139	153.82	5.15	7.57	31.489	08:14.4	19.88
7/25/2013	2139	140.59	5.23	7.99	31.492	08:18.4	19.67
7/26/2013	2139	127.36	5.31	8.41	31.495	08:22.3	19.45
7/27/2013	2139	114.12	5.40	8.82	31.499	08:26.2	19.23
7/28/2013	2139	100.89	5.48	9.23	31.502	08:30.2	19.00
7/29/2013	2139	87.72	5.55	9.64	31.505	08:34.1	18.77
7/30/2013	2139	74.49	5.63	10.05	31.509	08:38.0	18.53
7/31/2013	2139	61.26	5.71	10.45	31.513	08:41.9	18.28

(continued)

(continued)

Calendar	Rotation	Heliographic			Diameter	RA	Dec
Date	Number	*Lo*	*Bo*	*Po*	(arcmin)	(HH:MM)	(deg)
8/1/2013	2139	48.03	5.78	10.85	31.517	08:45.8	18.04
8/2/2013	2139	34.81	5.85	11.24	31.521	08:49.7	17.78
8/3/2013	2139	21.58	5.92	11.63	31.525	08:53.5	17.52
8/4/2013	2139	8.35	5.99	12.02	31.529	08:57.4	17.26
8/5/2013	2140	355.12	6.06	12.41	31.533	09:1.2	16.99
8/6/2013	2140	341.90	6.13	12.79	31.538	09:5.1	16.72
8/7/2013	2140	328.67	6.19	13.16	31.542	09:8.9	16.44
8/8/2013	2140	315.44	6.25	13.53	31.547	09:12.7	16.16
8/9/2013	2140	302.22	6.31	13.90	31.552	09:16.5	15.87
8/10/2013	2140	288.99	6.37	14.27	31.557	09:20.3	15.59
8/11/2013	2140	275.77	6.43	14.63	31.562	09:24.1	15.29
8/12/2013	2140	262.54	6.48	14.98	31.568	09:27.9	14.99
8/13/2013	2140	249.32	6.54	15.33	31.573	09:31.6	14.69
8/14/2013	2140	236.10	6.59	15.68	31.579	09:35.4	14.38
8/15/2013	2140	222.88	6.64	16.02	31.584	09:39.2	14.07
8/16/2013	2140	209.66	6.69	16.36	31.590	09:42.9	13.76
8/17/2013	2140	196.44	6.73	16.69	31.596	09:46.6	13.44
8/18/2013	2140	183.28	6.78	17.02	31.602	09:50.4	13.12
8/19/2013	2140	170.06	6.82	17.34	31.608	09:54.1	12.80
8/20/2013	2140	156.84	6.86	17.66	31.614	09:57.8	12.47
8/21/2013	2140	143.62	6.90	17.97	31.621	10:1.5	12.14
8/22/2013	2140	130.40	6.94	18.28	31.627	10:5.2	11.81
8/23/2013	2140	117.19	6.97	18.58	31.634	10:8.9	11.47
8/24/2013	2140	103.97	7.00	18.88	31.640	10:12.5	11.13
8/25/2013	2140	90.75	7.03	19.18	31.647	10:16.2	10.78
8/26/2013	2140	77.54	7.06	19.46	31.654	10:19.9	10.44
8/27/2013	2140	64.32	7.09	19.75	31.661	10:23.5	10.09
8/28/2013	2140	51.11	7.11	20.02	31.668	10:27.2	9.74
8/29/2013	2140	37.89	7.14	20.30	31.675	10:30.8	9.38
8/30/2013	2140	24.68	7.16	20.56	31.682	10:34.5	9.03
8/31/2013	2140	11.47	7.18	20.82	31.690	10:38.1	8.67
9/1/2013	2141	358.26	7.19	21.08	31.697	10:41.8	8.31
9/2/2013	2141	345.04	7.21	21.33	31.705	10:45.4	7.94
9/3/2013	2141	331.83	7.22	21.57	31.712	10:49.0	7.58
9/4/2013	2141	318.62	7.23	21.81	31.720	10:52.6	7.21
9/5/2013	2141	305.41	7.24	22.05	31.728	10:56.2	6.84
9/6/2013	2141	292.20	7.24	22.28	31.736	10:59.8	6.47
9/7/2013	2141	279.05	7.25	22.50	31.744	11:3.4	6.09

(continued)

(continued)

Calendar Date	Rotation Number	Heliographic Lo	Bo	Po	Diameter (arcmin)	RA (HH:MM)	Dec (deg)
9/8/2013	2141	265.85	7.25	22.71	31.752	11:7.0	5.72
9/9/2013	2141	252.64	7.25	22.92	31.760	11:10.6	5.34
9/10/2013	2141	239.43	7.25	23.13	31.768	11:14.2	4.96
9/11/2013	2141	226.22	7.24	23.33	31.776	11:17.8	4.58
9/12/2013	2141	213.02	7.24	23.52	31.784	11:21.4	4.20
9/13/2013	2141	199.81	7.23	23.71	31.793	11:25.0	3.82
9/14/2013	2141	186.61	7.22	23.89	31.801	11:28.6	3.44
9/15/2013	2141	173.40	7.21	24.06	31.810	11:32.2	3.05
9/16/2013	2141	160.20	7.19	24.23	31.818	11:35.8	2.67
9/17/2013	2141	146.99	7.17	24.39	31.827	11:39.4	2.28
9/18/2013	2141	133.79	7.15	24.54	31.836	11:43.0	1.89
9/19/2013	2141	120.59	7.13	24.69	31.844	11:46.6	1.51
9/20/2013	2141	107.38	7.11	24.84	31.853	11:50.1	1.12
9/21/2013	2141	94.18	7.09	24.97	31.862	11:53.7	0.73
9/22/2013	2141	80.98	7.06	25.10	31.871	11:57.3	0.34
9/23/2013	2141	67.78	7.03	25.23	31.880	12:0.9	−0.05
9/24/2013	2141	54.58	7.00	25.34	31.889	12:4.5	−0.44
9/25/2013	2141	41.38	6.96	25.45	31.898	12:8.1	−0.83
9/26/2013	2141	28.18	6.93	25.56	31.907	12:11.7	−1.22
9/27/2013	2141	14.98	6.89	25.65	31.916	12:15.3	−1.61
9/28/2013	2141	1.84	6.85	25.74	31.925	12:18.9	−2.00
9/29/2013	2142	348.64	6.81	25.82	31.934	12:22.5	−2.39
9/30/2013	2142	335.44	6.76	25.90	31.943	12:26.1	−2.78
10/1/2013	2142	322.25	6.72	25.97	31.952	12:29.8	−3.17
10/2/2013	2142	309.05	6.67	26.03	31.961	12:33.4	−3.55
10/3/2013	2142	295.85	6.62	26.09	31.970	12:37.0	−3.94
10/4/2013	2142	282.65	6.57	26.14	31.980	12:40.7	−4.33
10/5/2013	2142	269.46	6.51	26.18	31.989	12:44.3	−4.71
10/6/2013	2142	256.26	6.46	26.21	31.998	12:47.9	−5.10
10/7/2013	2142	243.06	6.40	26.24	32.007	12:51.6	−5.48
10/8/2013	2142	229.87	6.34	26.26	32.016	12:55.3	−5.86
10/9/2013	2142	216.67	6.28	26.27	32.026	12:58.9	−6.25
10/10/2013	2142	203.48	6.22	26.27	32.035	13:2.6	−6.63
10/11/2013	2142	190.28	6.15	26.27	32.044	13:6.3	−7.00
10/12/2013	2142	177.09	6.08	26.26	32.053	13:10.0	−7.38
10/13/2013	2142	163.90	6.01	26.25	32.062	13:13.7	−7.76
10/14/2013	2142	150.70	5.94	26.22	32.072	13:17.4	−8.13
10/15/2013	2142	137.51	5.87	26.19	32.081	13:21.1	−8.50

(continued)

(continued)

Calendar Date	Rotation Number	Heliographic *Lo*	*Bo*	*Po*	Diameter (arcmin)	RA (HH:MM)	Dec (deg)
10/16/2013	2142	124.32	5.80	26.15	32.090	13:24.8	−8.87
10/17/2013	2142	111.12	5.72	26.10	32.099	13:28.6	−9.24
10/18/2013	2142	97.99	5.64	26.05	32.108	13:32.3	−9.60
10/19/2013	2142	84.80	5.56	25.99	32.117	13:36.1	−9.97
10/20/2013	2142	71.61	5.48	25.92	32.126	13:39.8	−10.33
10/21/2013	2142	58.42	5.40	25.84	32.135	13:43.6	−10.68
10/22/2013	2142	45.23	5.31	25.75	32.144	13:47.4	−11.04
10/23/2013	2142	32.03	5.23	25.66	32.153	13:51.2	−11.39
10/24/2013	2142	18.84	5.14	25.56	32.162	13:55.0	−11.74
10/25/2013	2142	5.65	5.05	25.45	32.171	13:58.8	−12.09
10/26/2013	2143	352.46	4.96	25.33	32.179	14:2.7	−12.43
10/27/2013	2143	339.27	4.86	25.21	32.188	14:6.5	−12.77
10/28/2013	2143	326.08	4.77	25.08	32.197	14:10.4	−13.11
10/29/2013	2143	312.89	4.67	24.94	32.205	14:14.2	−13.44
10/30/2013	2143	299.70	4.58	24.79	32.214	14:18.1	−13.77
10/31/2013	2143	286.52	4.48	24.63	32.222	14:22.0	−14.10
11/1/2013	2143	273.33	4.38	24.47	32.231	14:25.9	−14.42
11/2/2013	2143	260.14	4.28	24.30	32.239	14:29.8	−14.74
11/3/2013	2143	246.95	4.17	24.12	32.248	14:33.8	−15.05
11/4/2013	2143	233.76	4.07	23.93	32.256	14:37.7	−15.36
11/5/2013	2143	220.57	3.96	23.74	32.264	14:41.7	−15.67
11/6/2013	2143	207.39	3.86	23.53	32.272	14:45.7	−15.97
11/7/2013	2143	194.26	3.75	23.32	32.280	14:49.7	−16.27
11/8/2013	2143	181.07	3.64	23.10	32.288	14:53.7	−16.56
11/9/2013	2143	167.89	3.53	22.88	32.296	14:57.7	−16.85
11/10/2013	2143	154.70	3.42	22.64	32.303	15:1.7	−17.13
11/11/2013	2143	141.51	3.30	22.40	32.311	15:5.8	−17.41
11/12/2013	2143	128.33	3.19	22.15	32.319	15:9.8	−17.69
11/13/2013	2143	115.14	3.08	21.89	32.326	15:13.9	−17.96
11/14/2013	2143	101.95	2.96	21.63	32.333	15:18.0	−18.22
11/15/2013	2143	88.77	2.84	21.36	32.341	15:22.1	−18.48
11/16/2013	2143	75.58	2.73	21.08	32.348	15:26.2	−18.73
11/17/2013	2143	62.40	2.61	20.79	32.355	15:30.3	−18.98
11/18/2013	2143	49.21	2.49	20.50	32.362	15:34.5	−19.22
11/19/2013	2143	36.03	2.37	20.19	32.369	15:38.6	−19.45
11/20/2013	2143	22.84	2.25	19.89	32.375	15:42.8	−19.68
11/21/2013	2143	9.66	2.13	19.57	32.382	15:47.0	−19.91
11/22/2013	2144	356.47	2.00	19.25	32.388	15:51.2	−20.13

(continued)

(continued)

Calendar	Rotation	Heliographic			Diameter	RA	Dec
Date	Number	*Lo*	*Bo*	*Po*	(arcmin)	(HH:MM)	(deg)
11/23/2013	2144	343.29	1.88	18.92	32.395	15:55.4	−20.34
11/24/2013	2144	330.11	1.76	18.58	32.401	15:59.6	−20.54
11/25/2013	2144	316.92	1.63	18.24	32.407	16:3.9	−20.74
11/26/2013	2144	303.74	1.51	17.88	32.413	16:8.1	−20.94
11/27/2013	2144	290.62	1.38	17.53	32.419	16:12.4	−21.12
11/28/2013	2144	277.44	1.26	17.16	32.425	16:16.7	−21.30
11/29/2013	2144	264.25	1.13	16.79	32.430	16:21.0	−21.47
11/30/2013	2144	251.07	1.01	16.42	32.436	16:25.3	−21.64
12/1/2013	2144	237.89	0.88	16.04	32.441	16:29.6	−21.80
12/2/2013	2144	224.71	0.75	15.65	32.446	16:33.9	−21.95
12/3/2013	2144	211.53	0.62	15.25	32.451	16:38.2	−22.10
12/4/2013	2144	198.34	0.50	14.86	32.456	16:42.5	−22.23
12/5/2013	2144	185.16	0.37	14.45	32.461	16:46.9	−22.37
12/6/2013	2144	171.98	0.24	14.04	32.466	16:51.3	−22.49
12/7/2013	2144	158.80	0.11	13.62	32.470	16:55.6	−22.61
12/8/2013	2144	145.62	−0.02	13.20	32.474	16:60.0	−22.71
12/9/2013	2144	132.44	−0.14	12.78	32.479	17:4.4	−22.82
12/10/2013	2144	119.26	−0.27	12.35	32.483	17:8.8	−22.91
12/11/2013	2144	106.08	−0.40	11.91	32.487	17:13.2	−23.00
12/12/2013	2144	92.90	−0.53	11.47	32.490	17:17.6	−23.08
12/13/2013	2144	79.72	−0.66	11.03	32.494	17:22.0	−23.15
12/14/2013	2144	66.54	−0.78	10.58	32.497	17:26.4	−23.21
12/15/2013	2144	53.36	−0.91	10.13	32.500	17:30.8	−23.27
12/16/2013	2144	40.19	−1.04	9.68	32.504	17:35.2	−23.32
12/17/2013	2144	27.07	−1.16	9.22	32.507	17:39.7	−23.36
12/18/2013	2144	13.89	−1.29	8.76	32.509	17:44.1	−23.39
12/19/2013	2144	0.71	−1.42	8.30	32.512	17:48.5	−23.41
12/20/2013	2145	347.54	−1.54	7.83	32.514	17:53.0	−23.43
12/21/2013	2145	334.36	−1.67	7.36	32.517	17:57.4	−23.44
12/22/2013	2145	321.18	−1.79	6.89	32.519	18:1.8	−23.44
12/23/2013	2145	308.01	−1.92	6.41	32.521	18:6.3	−23.44
12/24/2013	2145	294.83	−2.04	5.94	32.523	18:10.7	−23.42
12/25/2013	2145	281.65	−2.16	5.46	32.524	18:15.1	−23.40
12/26/2013	2145	268.48	−2.29	4.98	32.526	18:19.6	−23.37
12/27/2013	2145	255.30	−2.41	4.50	32.527	18:24.0	−23.33
12/28/2013	2145	242.13	−2.53	4.02	32.528	18:28.4	−23.29
12/29/2013	2145	228.95	−2.65	3.54	32.529	18:32.9	−23.23
12/30/2013	2145	215.78	−2.77	3.05	32.530	18:37.3	−23.17

(continued)

(continued)

Calendar	Rotation	Heliographic			Diameter	RA	Dec
Date	Number	*Lo*	*Bo*	*Po*	(arcmin)	(HH:MM)	(deg)
12/31/2013	2145	202.60	−2.89	2.57	32.530	18:41.7	−23.10
1/1/2014	2145	189.43	−3.00	2.08	32.531	18:46.1	−23.03
1/2/2014	2145	176.25	−3.12	1.60	32.531	18:50.5	−22.94
1/3/2014	2145	163.08	−3.24	1.11	32.531	18:54.9	−22.85
1/4/2014	2145	149.90	−3.35	0.63	32.531	18:59.3	−22.75
1/5/2014	2145	136.73	−3.46	0.15	32.531	19:3.7	−22.64
1/6/2014	2145	123.62	−3.58	−0.34	32.531	19:8.1	−22.53
1/7/2014	2145	110.45	−3.69	−0.82	32.530	19:12.5	−22.41
1/8/2014	2145	97.28	−3.80	−1.30	32.529	19:16.9	−22.28
1/9/2014	2145	84.10	−3.91	−1.78	32.529	19:21.2	−22.14
1/10/2014	2145	70.93	−4.02	−2.26	32.528	19:25.6	−22.00
1/11/2014	2145	57.76	−4.12	−2.74	32.526	19:29.9	−21.85
1/12/2014	2145	44.59	−4.23	−3.21	32.525	19:34.3	−21.69
1/13/2014	2145	31.41	−4.33	−3.69	32.523	19:38.6	−21.53
1/14/2014	2145	18.24	−4.43	−4.16	32.522	19:42.9	−21.36
1/15/2014	2145	5.07	−4.54	−4.63	32.520	19:47.2	−21.18
1/16/2014	2146	351.90	−4.63	−5.10	32.518	19:51.5	−20.99
1/17/2014	2146	338.73	−4.73	−5.56	32.515	19:55.8	−20.80
1/18/2014	2146	325.56	−4.83	−6.02	32.513	19:60.0	−20.60
1/19/2014	2146	312.39	−4.92	−6.48	32.511	20:4.3	−20.40
1/20/2014	2146	299.22	−5.02	−6.94	32.508	20:8.5	−20.19
1/21/2014	2146	286.05	−5.11	−7.39	32.505	20:12.8	−19.97
1/22/2014	2146	272.87	−5.20	−7.84	32.502	20:17.0	−19.74
1/23/2014	2146	259.70	−5.29	−8.29	32.499	20:21.2	−19.51
1/24/2014	2146	246.53	−5.38	−8.73	32.495	20:25.4	−19.28
1/25/2014	2146	233.36	−5.46	−9.17	32.492	20:29.6	−19.04
1/26/2014	2146	220.26	−5.55	−9.60	32.488	20:33.7	−18.79
1/27/2014	2146	207.09	−5.63	−10.04	32.484	20:37.9	−18.53
1/28/2014	2146	193.92	−5.71	−10.46	32.480	20:42.0	−18.28
1/29/2014	2146	180.75	−5.79	−10.89	32.476	20:46.2	−18.01
1/30/2014	2146	167.58	−5.86	−11.30	32.472	20:50.3	−17.74
1/31/2014	2146	154.41	−5.94	−11.72	32.468	20:54.4	−17.47
2/1/2014	2146	141.24	−6.01	−12.13	32.463	20:58.5	−17.18
2/2/2014	2146	128.07	−6.08	−12.54	32.458	21:2.5	−16.90
2/3/2014	2146	114.90	−6.15	−12.94	32.454	21:6.6	−16.61
2/4/2014	2146	101.73	−6.22	−13.33	32.449	21:10.6	−16.31
2/5/2014	2146	88.56	−6.28	−13.72	32.443	21:14.7	−16.01
2/6/2014	2146	75.39	−6.35	−14.11	32.438	21:18.7	−15.71

(continued)

(continued)

Calendar	Rotation	Heliographic			Diameter	RA	Dec
Date	Number	*Lo*	*Bo*	*Po*	(arcmin)	(HH:MM)	(deg)
2/7/2014	2146	62.22	−6.41	−14.49	32.433	21:22.7	−15.40
2/8/2014	2146	49.05	−6.47	−14.87	32.427	21:26.7	−15.09
2/9/2014	2146	35.88	−6.52	−15.24	32.422	21:30.7	−14.77
2/10/2014	2146	22.71	−6.58	−15.61	32.416	21:34.6	−14.45
2/11/2014	2146	9.54	−6.63	−15.97	32.410	21:38.6	−14.12
2/12/2014	2147	356.36	−6.68	−16.32	32.404	21:42.5	−13.79
2/13/2014	2147	343.19	−6.73	−16.67	32.398	21:46.4	−13.46
2/14/2014	2147	330.02	−6.78	−17.02	32.391	21:50.4	−13.12
2/15/2014	2147	316.85	−6.82	−17.36	32.385	21:54.3	−12.78
2/16/2014	2147	303.74	−6.86	−17.69	32.378	21:58.2	−12.44
2/17/2014	2147	290.57	−6.90	−18.02	32.372	22:2.0	−12.09
2/18/2014	2147	277.40	−6.94	−18.34	32.365	22:5.9	−11.74
2/19/2014	2147	264.23	−6.98	−18.66	32.358	22:9.8	−11.39
2/20/2014	2147	251.06	−7.01	−18.97	32.351	22:13.6	−11.03
2/21/2014	2147	237.88	−7.04	−19.27	32.344	22:17.4	−10.67
2/22/2014	2147	224.71	−7.07	−19.57	32.337	22:21.3	−10.31
2/23/2014	2147	211.54	−7.10	−19.86	32.329	22:25.1	−9.94
2/24/2014	2147	198.36	−7.12	−20.15	32.322	22:28.9	−9.57
2/25/2014	2147	185.19	−7.15	−20.43	32.315	22:32.7	−9.20
2/26/2014	2147	172.02	−7.17	−20.70	32.307	22:36.5	−8.83
2/27/2014	2147	158.84	−7.18	−20.97	32.299	22:40.2	−8.46
2/28/2014	2147	145.67	−7.20	−21.23	32.291	22:44.0	−8.08
3/1/2014	2147	132.49	−7.21	−21.49	32.284	22:47.8	−7.70
3/2/2014	2147	119.32	−7.23	−21.74	32.276	22:51.5	−7.32
3/3/2014	2147	106.14	−7.24	−21.98	32.268	22:55.2	−6.94
3/4/2014	2147	92.96	−7.24	−22.22	32.259	22:59.0	−6.56
3/5/2014	2147	79.79	−7.25	−22.45	32.251	23:2.7	−6.17
3/6/2014	2147	66.61	−7.25	−22.68	32.243	23:6.4	−5.78
3/7/2014	2147	53.43	−7.25	−22.89	32.235	23:10.1	−5.39
3/8/2014	2147	40.31	−7.25	−23.10	32.226	23:13.8	−5.01
3/9/2014	2147	27.14	−7.24	−23.31	32.218	23:17.5	−4.61
3/10/2014	2147	13.96	−7.24	−23.51	32.209	23:21.2	−4.22
3/11/2014	2148	0.77	−7.23	−23.70	32.201	23:24.9	−3.83
3/12/2014	2148	347.59	−7.22	−23.88	32.192	23:28.6	−3.44
3/13/2014	2148	334.41	−7.21	−24.06	32.183	23:32.3	−3.04
3/14/2014	2148	321.23	−7.19	−24.23	32.175	23:35.9	−2.65
3/15/2014	2148	308.05	−7.17	−24.40	32.166	23:39.6	−2.26
3/16/2014	2148	294.87	−7.15	−24.56	32.157	23:43.3	−1.86

(continued)

(continued)

Calendar	Rotation	Heliographic			Diameter	RA	Dec
Date	Number	Lo	Bo	Po	(arcmin)	(HH:MM)	(deg)
3/17/2014	2148	281.68	−7.13	−24.71	32.148	23:46.9	−1.46
3/18/2014	2148	268.50	−7.11	−24.85	32.139	23:50.6	−1.07
3/19/2014	2148	255.31	−7.08	−24.99	32.130	23:54.2	−0.67
3/20/2014	2148	242.13	−7.05	−25.12	32.121	23:57.9	−0.28
3/21/2014	2148	228.94	−7.02	−25.25	32.112	00:1.5	0.12
3/22/2014	2148	215.75	−6.99	−25.36	32.103	00:5.2	0.51
3/23/2014	2148	202.57	−6.96	−25.47	32.094	00:8.8	0.91
3/24/2014	2148	189.38	−6.92	−25.58	32.085	00:12.5	1.30
3/25/2014	2148	176.19	−6.88	−25.67	32.076	00:16.1	1.70
3/26/2014	2148	163.00	−6.84	−25.76	32.067	00:19.8	2.09
3/27/2014	2148	149.81	−6.80	−25.84	32.057	00:23.4	2.48
3/28/2014	2148	136.68	−6.75	−25.92	32.048	00:27.0	2.87
3/29/2014	2148	123.49	−6.71	−25.99	32.039	00:30.7	3.26
3/30/2014	2148	110.29	−6.66	−26.05	32.030	00:34.3	3.65
3/31/2014	2148	97.10	−6.61	−26.10	32.021	00:38.0	4.04
4/1/2014	2148	83.91	−6.56	−26.15	32.011	00:41.6	4.43
4/2/2014	2148	70.71	−6.50	−26.19	32.002	00:45.3	4.82
4/3/2014	2148	57.51	−6.44	−26.22	31.993	00:48.9	5.20
4/4/2014	2148	44.32	−6.39	−26.24	31.984	00:52.6	5.58
4/5/2014	2148	31.12	−6.33	−26.26	31.974	00:56.2	5.96
4/6/2014	2148	17.92	−6.26	−26.27	31.965	00:59.9	6.34
4/7/2014	2148	4.72	−6.20	−26.27	31.956	01:3.5	6.72
4/8/2014	2149	351.52	−6.14	−26.27	31.947	01:7.2	7.10
4/9/2014	2149	338.32	−6.07	−26.26	31.938	01:10.9	7.47
4/10/2014	2149	325.12	−6.00	−26.24	31.929	01:14.5	7.84
4/11/2014	2149	311.92	−5.93	−26.21	31.920	01:18.2	8.21
4/12/2014	2149	298.71	−5.86	−26.18	31.911	01:21.9	8.58
4/13/2014	2149	285.51	−5.78	−26.14	31.902	01:25.6	8.94
4/14/2014	2149	272.30	−5.71	−26.09	31.893	01:29.3	9.31
4/15/2014	2149	259.10	−5.63	−26.04	31.884	01:33.0	9.67
4/16/2014	2149	245.89	−5.55	−25.97	31.875	01:36.7	10.02
4/17/2014	2149	232.75	−5.47	−25.90	31.866	01:40.4	10.38
4/18/2014	2149	219.54	−5.39	−25.83	31.857	01:44.1	10.73
4/19/2014	2149	206.33	−5.30	−25.74	31.848	01:47.8	11.08
4/20/2014	2149	193.12	−5.22	−25.65	31.839	01:51.5	11.42
4/21/2014	2149	179.91	−5.13	−25.55	31.831	01:55.3	11.77
4/22/2014	2149	166.70	−5.04	−25.44	31.822	01:59.0	12.10
4/23/2014	2149	153.49	−4.95	−25.33	31.814	02:2.8	12.44

(continued)

(continued)

Calendar	Rotation	Heliographic			Diameter	RA	Dec
Date	Number	*Lo*	*Bo*	*Po*	(arcmin)	(HH:MM)	(deg)
4/24/2014	2149	140.27	−4.86	−25.21	31.805	02:6.5	12.77
4/25/2014	2149	127.06	−4.77	−25.08	31.797	02:10.3	13.10
4/26/2014	2149	113.84	−4.68	−24.94	31.788	02:14.1	13.43
4/27/2014	2149	100.63	−4.58	−24.80	31.780	02:17.9	13.75
4/28/2014	2149	87.41	−4.49	−24.65	31.772	02:21.6	14.07
4/29/2014	2149	74.19	−4.39	−24.49	31.763	02:25.4	14.38
4/30/2014	2149	60.98	−4.29	−24.32	31.755	02:29.2	14.69
5/1/2014	2149	47.76	−4.19	−24.15	31.747	02:33.1	15.00
5/2/2014	2149	34.54	−4.09	−23.97	31.739	02:36.9	15.30
5/3/2014	2149	21.32	−3.99	−23.78	31.731	02:40.7	15.59
5/4/2014	2149	8.10	−3.89	−23.59	31.723	02:44.6	15.89
5/5/2014	2150	354.88	−3.78	−23.39	31.716	02:48.4	16.18
5/6/2014	2150	341.65	−3.68	−23.18	31.708	02:52.3	16.46
5/7/2014	2150	328.49	−3.57	−22.96	31.701	02:56.1	16.74
5/8/2014	2150	315.27	−3.46	−22.74	31.693	02:60.0	17.02
5/9/2014	2150	302.05	−3.36	−22.51	31.686	03:3.9	17.29
5/10/2014	2150	288.82	−3.25	−22.28	31.678	03:7.8	17.55
5/11/2014	2150	275.60	−3.14	−22.03	31.671	03:11.7	17.81
5/12/2014	2150	262.37	−3.03	−21.78	31.664	03:15.6	18.07
5/13/2014	2150	249.14	−2.92	−21.53	31.657	03:19.6	18.32
5/14/2014	2150	235.91	−2.80	−21.26	31.650	03:23.5	18.56
5/15/2014	2150	222.69	−2.69	−20.99	31.643	03:27.4	18.80
5/16/2014	2150	209.46	−2.58	−20.72	31.637	03:31.4	19.04
5/17/2014	2150	196.23	−2.46	−20.43	31.630	03:35.4	19.27
5/18/2014	2150	183.00	−2.35	−20.14	31.624	03:39.3	19.49
5/19/2014	2150	169.77	−2.23	−19.85	31.617	03:43.3	19.71
5/20/2014	2150	156.54	−2.12	−19.55	31.611	03:47.3	19.92
5/21/2014	2150	143.30	−2.00	−19.24	31.605	03:51.3	20.13
5/22/2014	2150	130.07	−1.88	−18.92	31.599	03:55.3	20.33
5/23/2014	2150	116.84	−1.77	−18.60	31.593	03:59.3	20.53
5/24/2014	2150	103.61	−1.65	−18.28	31.587	04:3.4	20.72
5/25/2014	2150	90.37	−1.53	−17.95	31.581	04:7.4	20.90
5/26/2014	2150	77.14	−1.41	−17.61	31.576	04:11.4	21.08
5/27/2014	2150	63.97	−1.29	−17.26	31.570	04:15.5	21.25
5/28/2014	2150	50.73	−1.17	−16.92	31.565	04:19.5	21.42
5/29/2014	2150	37.49	−1.05	−16.56	31.560	04:23.6	21.58
5/30/2014	2150	24.26	−0.93	−16.20	31.554	04:27.7	21.73
5/31/2014	2150	11.02	−0.81	−15.84	31.549	04:31.8	21.88

(continued)

(continued)

Calendar	Rotation	Heliographic			Diameter	RA	Dec
Date	Number	*Lo*	*Bo*	*Po*	(arcmin)	(HH:MM)	(deg)
6/1/2014	2151	357.78	−0.69	−15.47	31.545	04:35.9	22.02
6/2/2014	2151	344.55	−0.57	−15.09	31.540	04:40.0	22.15
6/3/2014	2151	331.31	−0.45	−14.72	31.535	04:44.1	22.28
6/4/2014	2151	318.07	−0.33	−14.33	31.531	04:48.2	22.40
6/5/2014	2151	304.83	−0.21	−13.94	31.527	04:52.3	22.52
6/6/2014	2151	291.60	−0.09	−13.55	31.522	04:56.4	22.63
6/7/2014	2151	278.36	0.03	−13.15	31.518	05:0.5	22.73
6/8/2014	2151	265.12	0.15	−12.75	31.514	05:4.6	22.82
6/9/2014	2151	251.88	0.27	−12.35	31.511	05:8.8	22.91
6/10/2014	2151	238.64	0.39	−11.94	31.507	05:12.9	22.99
6/11/2014	2151	225.40	0.51	−11.53	31.503	05:17.0	23.07
6/12/2014	2151	212.16	0.63	−11.11	31.500	05:21.2	23.14
6/13/2014	2151	198.92	0.75	−10.69	31.497	05:25.3	23.20
6/14/2014	2151	185.68	0.87	−10.27	31.494	05:29.5	23.25
6/15/2014	2151	172.44	0.99	−9.84	31.491	05:33.6	23.30
6/16/2014	2151	159.26	1.11	−9.41	31.488	05:37.8	23.34
6/17/2014	2151	146.02	1.23	−8.98	31.485	05:42.0	23.37
6/18/2014	2151	132.78	1.35	−8.55	31.483	05:46.1	23.40
6/19/2014	2151	119.54	1.47	−8.11	31.480	05:50.3	23.42
6/20/2014	2151	106.30	1.58	−7.68	31.478	05:54.4	23.43
6/21/2014	2151	93.06	1.70	−7.23	31.476	05:58.6	23.44
6/22/2014	2151	79.82	1.82	−6.79	31.474	06:2.7	23.44
6/23/2014	2151	66.58	1.93	−6.35	31.472	06:6.9	23.43
6/24/2014	2151	53.34	2.05	−5.90	31.471	06:11.0	23.42
6/25/2014	2151	40.10	2.16	−5.45	31.469	06:15.2	23.40
6/26/2014	2151	26.86	2.28	−5.01	31.468	06:19.4	23.37
6/27/2014	2151	13.62	2.39	−4.56	31.466	06:23.5	23.34
6/28/2014	2151	0.38	2.51	−4.11	31.465	06:27.7	23.29
6/29/2014	2152	347.14	2.62	−3.65	31.464	06:31.8	23.25
6/30/2014	2152	333.89	2.73	−3.20	31.464	06:35.9	23.19
7/1/2014	2152	320.65	2.84	−2.75	31.463	06:40.1	23.13
7/2/2014	2152	307.42	2.95	−2.30	31.462	06:44.2	23.06
7/3/2014	2152	294.18	3.06	−1.84	31.462	06:48.3	22.99
7/4/2014	2152	280.94	3.17	−1.39	31.462	06:52.5	22.90
7/5/2014	2152	267.70	3.28	−0.94	31.462	06:56.6	22.82
7/6/2014	2152	254.46	3.39	−0.48	31.462	07:0.7	22.72
7/7/2014	2152	241.28	3.49	−0.03	31.462	07:4.8	22.62
7/8/2014	2152	228.04	3.60	0.42	31.463	07:8.9	22.51

(continued)

(continued)

Calendar	Rotation	Heliographic			Diameter	RA	Dec
Date	Number	*Lo*	*Bo*	*Po*	(arcmin)	(HH:MM)	(deg)
7/9/2014	2152	214.80	3.70	0.87	31.463	07:13.0	22.40
7/10/2014	2152	201.56	3.80	1.32	31.464	07:17.1	22.27
7/11/2014	2152	188.33	3.91	1.77	31.465	07:21.2	22.15
7/12/2014	2152	175.09	4.01	2.22	31.466	07:25.2	22.01
7/13/2014	2152	161.85	4.11	2.67	31.467	07:29.3	21.87
7/14/2014	2152	148.61	4.20	3.11	31.468	07:33.3	21.73
7/15/2014	2152	135.38	4.30	3.56	31.469	07:37.4	21.57
7/16/2014	2152	122.14	4.40	4.00	31.471	07:41.4	21.41
7/17/2014	2152	108.90	4.49	4.44	31.473	07:45.5	21.25
7/18/2014	2152	95.67	4.59	4.88	31.475	07:49.5	21.08
7/19/2014	2152	82.43	4.68	5.32	31.477	07:53.5	20.90
7/20/2014	2152	69.20	4.77	5.75	31.479	07:57.5	20.72
7/21/2014	2152	55.96	4.86	6.18	31.481	08:1.5	20.53
7/22/2014	2152	42.73	4.95	6.61	31.483	08:5.5	20.34
7/23/2014	2152	29.50	5.04	7.04	31.486	08:9.5	20.14
7/24/2014	2152	16.26	5.12	7.46	31.489	08:13.4	19.93
7/25/2014	2152	3.03	5.21	7.88	31.492	08:17.4	19.72
7/26/2014	2153	349.80	5.29	8.30	31.495	08:21.4	19.50
7/27/2014	2153	336.63	5.37	8.72	31.498	08:25.3	19.28
7/28/2014	2153	323.40	5.45	9.13	31.501	08:29.2	19.06
7/29/2014	2153	310.17	5.53	9.54	31.504	08:33.1	18.82
7/30/2014	2153	296.94	5.61	9.95	31.508	08:37.0	18.59
7/31/2014	2153	283.70	5.69	10.35	31.512	08:40.9	18.34
8/1/2014	2153	270.48	5.76	10.75	31.516	08:44.8	18.10
8/2/2014	2153	257.25	5.83	11.15	31.520	08:48.7	17.84
8/3/2014	2153	244.02	5.91	11.54	31.524	08:52.6	17.59
8/4/2014	2153	230.79	5.97	11.93	31.528	08:56.4	17.32
8/5/2014	2153	217.56	6.04	12.31	31.532	08:60.3	17.06
8/6/2014	2153	204.34	6.11	12.69	31.537	09:4.1	16.79
8/7/2014	2153	191.11	6.17	13.07	31.541	09:8.0	16.51
8/8/2014	2153	177.88	6.24	13.44	31.546	09:11.8	16.23
8/9/2014	2153	164.66	6.30	13.81	31.551	09:15.6	15.94
8/10/2014	2153	151.43	6.36	14.18	31.556	09:19.4	15.66
8/11/2014	2153	138.21	6.41	14.54	31.561	09:23.2	15.36
8/12/2014	2153	124.99	6.47	14.89	31.566	09:27.0	15.07
8/13/2014	2153	111.76	6.52	15.25	31.572	09:30.7	14.76
8/14/2014	2153	98.54	6.58	15.59	31.577	09:34.5	14.46
8/15/2014	2153	85.32	6.63	15.94	31.583	09:38.2	14.15

(continued)

(continued)

Calendar	Rotation	Heliographic			Diameter	RA	Dec
Date	Number	*Lo*	*Bo*	*Po*	(arcmin)	(HH:MM)	(deg)
8/16/2014	2153	72.16	6.67	16.27	31.589	09:42.0	13.84
8/17/2014	2153	58.94	6.72	16.61	31.594	09:45.7	13.52
8/18/2014	2153	45.72	6.77	16.94	31.600	09:49.5	13.20
8/19/2014	2153	32.50	6.81	17.26	31.607	09:53.2	12.88
8/20/2014	2153	19.28	6.85	17.58	31.613	09:56.9	12.55
8/21/2014	2153	6.06	6.89	17.89	31.619	10:0.6	12.22
8/22/2014	2154	352.84	6.93	18.20	31.625	10:4.3	11.89
8/23/2014	2154	339.62	6.96	18.51	31.632	10:8.0	11.55
8/24/2014	2154	326.41	6.99	18.81	31.639	10:11.7	11.21
8/25/2014	2154	313.19	7.03	19.10	31.645	10:15.3	10.87
8/26/2014	2154	299.97	7.05	19.39	31.652	10:19.0	10.52
8/27/2014	2154	286.76	7.08	19.68	31.659	10:22.7	10.17
8/28/2014	2154	273.55	7.11	19.95	31.666	10:26.3	9.82
8/29/2014	2154	260.33	7.13	20.23	31.673	10:30.0	9.47
8/30/2014	2154	247.12	7.15	20.50	31.681	10:33.6	9.11
8/31/2014	2154	233.91	7.17	20.76	31.688	10:37.2	8.75
9/1/2014	2154	220.69	7.19	21.02	31.695	10:40.9	8.39
9/2/2014	2154	207.48	7.20	21.27	31.703	10:44.5	8.03
9/3/2014	2154	194.27	7.22	21.51	31.710	10:48.1	7.66
9/4/2014	2154	181.06	7.23	21.75	31.718	10:51.7	7.30
9/5/2014	2154	167.91	7.24	21.99	31.726	10:55.4	6.93
9/6/2014	2154	154.70	7.24	22.22	31.734	10:59.0	6.56
9/7/2014	2154	141.49	7.25	22.44	31.742	11:2.6	6.18
9/8/2014	2154	128.28	7.25	22.66	31.750	11:6.2	5.81
9/9/2014	2154	115.07	7.25	22.87	31.758	11:9.8	5.43
9/10/2014	2154	101.87	7.25	23.08	31.766	11:13.4	5.05
9/11/2014	2154	88.66	7.24	23.28	31.774	11:17.0	4.68
9/12/2014	2154	75.45	7.24	23.47	31.782	11:20.6	4.29
9/13/2014	2154	62.25	7.23	23.66	31.791	11:24.2	3.91
9/14/2014	2154	49.04	7.22	23.84	31.799	11:27.7	3.53
9/15/2014	2154	35.84	7.21	24.02	31.808	11:31.3	3.15
9/16/2014	2154	22.63	7.19	24.19	31.816	11:34.9	2.76
9/17/2014	2154	9.43	7.18	24.35	31.825	11:38.5	2.37
9/18/2014	2155	356.22	7.16	24.51	31.833	11:42.1	1.99
9/19/2014	2155	343.02	7.14	24.66	31.842	11:45.7	1.60
9/20/2014	2155	329.82	7.12	24.80	31.851	11:49.3	1.21
9/21/2014	2155	316.62	7.09	24.94	31.860	11:52.9	0.82
9/22/2014	2155	303.41	7.06	25.07	31.869	11:56.5	0.43

(continued)

(continued)

Calendar	Rotation	Heliographic			Diameter	RA	Dec
Date	Number	*Lo*	*Bo*	*Po*	(arcmin)	(HH:MM)	(deg)
9/23/2014	2155	290.21	7.04	25.19	31.877	11:60.0	0.04
9/24/2014	2155	277.01	7.01	25.31	31.886	12:3.6	−0.35
9/25/2014	2155	263.87	6.97	25.42	31.895	12:7.2	−0.74
9/26/2014	2155	250.67	6.94	25.53	31.904	12:10.8	−1.13
9/27/2014	2155	237.47	6.90	25.63	31.913	12:14.4	−1.52
9/28/2014	2155	224.27	6.86	25.72	31.922	12:18.0	−1.91
9/29/2014	2155	211.08	6.82	25.80	31.932	12:21.7	−2.29
9/30/2014	2155	197.88	6.78	25.88	31.941	12:25.3	−2.68
10/1/2014	2155	184.68	6.73	25.95	31.950	12:28.9	−3.07
10/2/2014	2155	171.48	6.68	26.02	31.959	12:32.5	−3.46
10/3/2014	2155	158.28	6.63	26.07	31.968	12:36.1	−3.85
10/4/2014	2155	145.09	6.58	26.12	31.977	12:39.8	−4.23
10/5/2014	2155	131.89	6.53	26.17	31.986	12:43.4	−4.62
10/6/2014	2155	118.69	6.47	26.20	31.996	12:47.1	−5.01
10/7/2014	2155	105.50	6.42	26.23	32.005	12:50.7	−5.39
10/8/2014	2155	92.30	6.36	26.25	32.014	12:54.4	−5.77
10/9/2014	2155	79.11	6.30	26.27	32.023	12:58.0	−6.15
10/10/2014	2155	65.91	6.23	26.27	32.033	13:1.7	−6.53
10/11/2014	2155	52.72	6.17	26.27	32.042	13:5.4	−6.91
10/12/2014	2155	39.52	6.10	26.26	32.051	13:9.1	−7.29
10/13/2014	2155	26.33	6.03	26.25	32.060	13:12.8	−7.67
10/14/2014	2155	13.14	5.96	26.23	32.069	13:16.5	−8.04
10/15/2014	2155	0.00	5.89	26.20	32.078	13:20.2	−8.41
10/16/2014	2156	346.81	5.82	26.16	32.088	13:23.9	−8.78
10/17/2014	2156	333.62	5.74	26.11	32.097	13:27.7	−9.15
10/18/2014	2156	320.43	5.66	26.06	32.106	13:31.4	−9.51
10/19/2014	2156	307.23	5.58	26.00	32.115	13:35.2	−9.88
10/20/2014	2156	294.04	5.50	25.93	32.124	13:38.9	−10.24
10/21/2014	2156	280.85	5.42	25.86	32.133	13:42.7	−10.60
10/22/2014	2156	267.66	5.33	25.77	32.142	13:46.5	−10.95
10/23/2014	2156	254.47	5.25	25.68	32.151	13:50.3	−11.31
10/24/2014	2156	241.28	5.16	25.58	32.160	13:54.1	−11.66
10/25/2014	2156	228.09	5.07	25.48	32.168	13:57.9	−12.00
10/26/2014	2156	214.90	4.98	25.36	32.177	14:1.7	−12.35
10/27/2014	2156	201.71	4.89	25.24	32.186	14:5.6	−12.69
10/28/2014	2156	188.52	4.79	25.11	32.195	14:9.4	−13.03
10/29/2014	2156	175.33	4.70	24.97	32.203	14:13.3	−13.36
10/30/2014	2156	162.14	4.60	24.82	32.212	14:17.2	−13.69

(continued)

(continued)

Calendar	Rotation	Heliographic			Diameter	RA	Dec
Date	Number	*Lo*	*Bo*	*Po*	(arcmin)	(HH:MM)	(deg)
10/31/2014	2156	148.95	4.50	24.67	32.220	14:21.1	−14.02
11/1/2014	2156	135.76	4.40	24.51	32.229	14:25.0	−14.34
11/2/2014	2156	122.57	4.30	24.34	32.237	14:28.9	−14.66
11/3/2014	2156	109.38	4.20	24.16	32.246	14:32.8	−14.98
11/4/2014	2156	96.25	4.09	23.98	32.254	14:36.8	−15.29
11/5/2014	2156	83.07	3.99	23.78	32.262	14:40.7	−15.60
11/6/2014	2156	69.88	3.88	23.58	32.270	14:44.7	−15.90
11/7/2014	2156	56.69	3.78	23.37	32.278	14:48.7	−16.20
11/8/2014	2156	43.50	3.67	23.16	32.286	14:52.7	−16.49
11/9/2014	2156	30.32	3.56	22.93	32.294	14:56.7	−16.78
11/10/2014	2156	17.13	3.45	22.70	32.302	15:0.7	−17.07
11/11/2014	2156	3.94	3.33	22.46	32.309	15:4.8	−17.35
11/12/2014	2157	350.76	3.22	22.21	32.317	15:8.8	−17.62
11/13/2014	2157	337.57	3.11	21.96	32.324	15:12.9	−17.89
11/14/2014	2157	324.38	2.99	21.69	32.332	15:17.0	−18.16
11/15/2014	2157	311.20	2.87	21.42	32.339	15:21.1	−18.42
11/16/2014	2157	298.01	2.76	21.15	32.346	15:25.2	−18.67
11/17/2014	2157	284.83	2.64	20.86	32.353	15:29.3	−18.92
11/18/2014	2157	271.64	2.52	20.57	32.360	15:33.5	−19.16
11/19/2014	2157	258.46	2.40	20.27	32.367	15:37.6	−19.40
11/20/2014	2157	245.27	2.28	19.96	32.374	15:41.8	−19.63
11/21/2014	2157	232.09	2.16	19.65	32.380	15:46.0	−19.85
11/22/2014	2157	218.90	2.04	19.32	32.387	15:50.2	−20.07
11/23/2014	2157	205.72	1.91	19.00	32.393	15:54.4	−20.29
11/24/2014	2157	192.54	1.79	18.66	32.399	15:58.6	−20.49
11/25/2014	2157	179.42	1.67	18.32	32.406	16:2.9	−20.69
11/26/2014	2157	166.23	1.54	17.97	32.412	16:7.1	−20.89
11/27/2014	2157	153.05	1.42	17.61	32.418	16:11.4	−21.08
11/28/2014	2157	139.87	1.29	17.25	32.423	16:15.6	−21.26
11/29/2014	2157	126.68	1.16	16.88	32.429	16:19.9	−21.43
11/30/2014	2157	113.50	1.04	16.51	32.434	16:24.2	−21.60
12/1/2014	2157	100.32	0.91	16.13	32.440	16:28.5	−21.76
12/2/2014	2157	87.14	0.78	15.74	32.445	16:32.8	−21.92
12/3/2014	2157	73.95	0.66	15.35	32.450	16:37.2	−22.06
12/4/2014	2157	60.77	0.53	14.95	32.455	16:41.5	−22.20
12/5/2014	2157	47.59	0.40	14.55	32.460	16:45.8	−22.33
12/6/2014	2157	34.41	0.27	14.14	32.465	16:50.2	−22.46
12/7/2014	2157	21.23	0.15	13.73	32.469	16:54.6	−22.58

(continued)

(continued)

Calendar	Rotation	Heliographic			Diameter	RA	Dec
Date	Number	*Lo*	*Bo*	*Po*	(arcmin)	(HH:MM)	(deg)
12/8/2014	2157	8.05	0.02	13.31	32.473	16:58.9	−22.69
12/9/2014	2158	354.87	−0.11	12.88	32.478	17:3.3	−22.79
12/10/2014	2158	341.69	−0.24	12.45	32.482	17:7.7	−22.89
12/11/2014	2158	328.51	−0.37	12.02	32.486	17:12.1	−22.98
12/12/2014	2158	315.33	−0.49	11.58	32.489	17:16.5	−23.06
12/13/2014	2158	302.15	−0.62	11.14	32.493	17:20.9	−23.13
12/14/2014	2158	288.97	−0.75	10.69	32.496	17:25.3	−23.20
12/15/2014	2158	275.86	−0.88	10.24	32.500	17:29.7	−23.25
12/16/2014	2158	262.68	−1.01	9.79	32.503	17:34.2	−23.30
12/17/2014	2158	249.50	−1.13	9.33	32.506	17:38.6	−23.35
12/18/2014	2158	236.32	−1.26	8.87	32.509	17:43.0	−23.38
12/19/2014	2158	223.14	−1.38	8.41	32.511	17:47.5	−23.41
12/20/2014	2158	209.97	−1.51	7.94	32.514	17:51.9	−23.43
12/21/2014	2158	196.79	−1.64	7.47	32.516	17:56.3	−23.44
12/22/2014	2158	183.61	−1.76	7.00	32.518	18:0.8	−23.44
12/23/2014	2158	170.43	−1.89	6.53	32.520	18:5.2	−23.44
12/24/2014	2158	157.26	−2.01	6.05	32.522	18:9.6	−23.43
12/25/2014	2158	144.08	−2.13	5.58	32.524	18:14.1	−23.40
12/26/2014	2158	130.91	−2.25	5.10	32.525	18:18.5	−23.38
12/27/2014	2158	117.73	−2.38	4.62	32.527	18:22.9	−23.34
12/28/2014	2158	104.55	−2.50	4.14	32.528	18:27.4	−23.30
12/29/2014	2158	91.38	−2.62	3.65	32.529	18:31.8	−23.25
12/30/2014	2158	78.20	−2.74	3.17	32.530	18:36.2	−23.19
12/31/2014	2158	65.03	−2.86	2.69	32.530	18:40.6	−23.12
1/1/2015	2158	51.86	−2.97	2.20	32.531	18:45.1	−23.05
1/2/2015	2158	38.68	−3.09	1.72	32.531	18:49.5	−22.96
1/3/2015	2158	25.51	−3.21	1.23	32.531	18:53.9	−22.87
1/4/2015	2158	12.40	−3.32	0.75	32.531	18:58.3	−22.78
1/5/2015	2159	359.22	−3.44	0.26	32.531	19:2.7	−22.67
1/6/2015	2159	346.05	−3.55	−0.22	32.531	19:7.1	−22.56
1/7/2015	2159	332.88	−3.66	−0.70	32.530	19:11.4	−22.44
1/8/2015	2159	319.70	−3.77	−1.18	32.530	19:15.8	−22.31
1/9/2015	2159	306.53	−3.88	−1.67	32.529	19:20.2	−22.18
1/10/2015	2159	293.36	−3.99	−2.14	32.528	19:24.5	−22.04
1/11/2015	2159	280.18	−4.10	−2.62	32.527	19:28.9	−21.89
1/12/2015	2159	267.01	−4.20	−3.10	32.525	19:33.2	−21.73
1/13/2015	2159	253.84	−4.30	−3.57	32.524	19:37.5	−21.57
1/14/2015	2159	240.67	−4.41	−4.05	32.522	19:41.9	−21.40

(continued)

(continued)

Calendar Date	Rotation Number	Heliographic Lo	Bo	Po	Diameter (arcmin)	RA (HH:MM)	Dec (deg)
1/15/2015	2159	227.50	−4.51	−4.51	32.520	19:46.2	−21.22
1/16/2015	2159	214.33	−4.61	−4.98	32.518	19:50.5	−21.04
1/17/2015	2159	201.16	−4.71	−5.45	32.516	19:54.7	−20.85
1/18/2015	2159	187.98	−4.81	−5.91	32.514	19:59.0	−20.65
1/19/2015	2159	174.81	−4.90	−6.37	32.511	20:3.3	−20.45
1/20/2015	2159	161.64	−4.99	−6.83	32.508	20:7.5	−20.24
1/21/2015	2159	148.47	−5.09	−7.28	32.506	20:11.7	−20.02
1/22/2015	2159	135.30	−5.18	−7.73	32.503	20:16.0	−19.80
1/23/2015	2159	122.13	−5.27	−8.18	32.500	20:20.2	−19.57
1/24/2015	2159	109.02	−5.35	−8.62	32.496	20:24.4	−19.34
1/25/2015	2159	95.85	−5.44	−9.06	32.493	20:28.6	−19.09
1/26/2015	2159	82.68	−5.53	−9.50	32.489	20:32.7	−18.85
1/27/2015	2159	69.51	−5.61	−9.93	32.485	20:36.9	−18.60
1/28/2015	2159	56.34	−5.69	−10.36	32.481	20:41.0	−18.34
1/29/2015	2159	43.17	−5.77	−10.78	32.477	20:45.2	−18.08
1/30/2015	2159	30.00	−5.84	−11.20	32.473	20:49.3	−17.81
1/31/2015	2159	16.83	−5.92	−11.62	32.469	20:53.4	−17.53
2/1/2015	2159	3.66	−5.99	−12.03	32.464	20:57.5	−17.25
2/2/2015	2160	350.49	−6.06	−12.44	32.460	21:1.5	−16.97
2/3/2015	2160	337.32	−6.13	−12.84	32.455	21:5.6	−16.68
2/4/2015	2160	324.15	−6.20	−13.24	32.450	21:9.7	−16.39
2/5/2015	2160	310.98	−6.27	−13.63	32.445	21:13.7	−16.09
2/6/2015	2160	297.81	−6.33	−14.02	32.439	21:17.7	−15.78
2/7/2015	2160	284.64	−6.39	−14.40	32.434	21:21.7	−15.48
2/8/2015	2160	271.47	−6.45	−14.78	32.429	21:25.7	−15.16
2/9/2015	2160	258.30	−6.51	−15.15	32.423	21:29.7	−14.85
2/10/2015	2160	245.13	−6.56	−15.52	32.417	21:33.7	−14.53
2/11/2015	2160	231.96	−6.62	−15.88	32.411	21:37.6	−14.20
2/12/2015	2160	218.79	−6.67	−16.24	32.405	21:41.6	−13.87
2/13/2015	2160	205.68	−6.72	−16.59	32.399	21:45.5	−13.54
2/14/2015	2160	192.51	−6.77	−16.93	32.393	21:49.4	−13.20
2/15/2015	2160	179.34	−6.81	−17.27	32.386	21:53.3	−12.86
2/16/2015	2160	166.17	−6.85	−17.61	32.380	21:57.2	−12.52
2/17/2015	2160	153.00	−6.89	−17.94	32.373	22:1.1	−12.17
2/18/2015	2160	139.83	−6.93	−18.26	32.367	22:5.0	−11.82
2/19/2015	2160	126.66	−6.97	−18.58	32.360	22:8.8	−11.47
2/20/2015	2160	113.48	−7.00	−18.89	32.353	22:12.7	−11.12
2/21/2015	2160	100.31	−7.03	−19.20	32.346	22:16.5	−10.76

(continued)

(continued)

Calendar Date	Rotation Number	Heliographic Lo	Bo	Po	Diameter (arcmin)	RA (HH:MM)	Dec (deg)
2/22/2015	2160	87.14	−7.06	−19.50	32.338	22:20.3	−10.39
2/23/2015	2160	73.96	−7.09	−19.79	32.331	22:24.2	−10.03
2/24/2015	2160	60.79	−7.12	−20.08	32.324	22:28.0	−9.66
2/25/2015	2160	47.62	−7.14	−20.36	32.316	22:31.8	−9.29
2/26/2015	2160	34.44	−7.16	−20.64	32.309	22:35.5	−8.92
2/27/2015	2160	21.27	−7.18	−20.91	32.301	22:39.3	−8.55
2/28/2015	2160	8.10	−7.20	−21.17	32.293	22:43.1	−8.17
3/1/2015	2161	354.92	−7.21	−21.43	32.285	22:46.9	−7.79
3/2/2015	2161	341.74	−7.22	−21.68	32.278	22:50.6	−7.41
3/3/2015	2161	328.57	−7.23	−21.92	32.270	22:54.3	−7.03
3/4/2015	2161	315.39	−7.24	−22.16	32.261	22:58.1	−6.65
3/5/2015	2161	302.28	−7.25	−22.39	32.253	23:1.8	−6.26
3/6/2015	2161	289.10	−7.25	−22.62	32.245	23:5.5	−5.88
3/7/2015	2161	275.92	−7.25	−22.84	32.237	23:9.2	−5.49
3/8/2015	2161	262.74	−7.25	−23.05	32.228	23:12.9	−5.10
3/9/2015	2161	249.56	−7.25	−23.26	32.220	23:16.6	−4.71
3/10/2015	2161	236.38	−7.24	−23.46	32.211	23:20.3	−4.32
3/11/2015	2161	223.20	−7.23	−23.65	32.203	23:24.0	−3.93
3/12/2015	2161	210.02	−7.22	−23.84	32.194	23:27.7	−3.53
3/13/2015	2161	196.84	−7.21	−24.02	32.185	23:31.4	−3.14
3/14/2015	2161	183.66	−7.19	−24.19	32.177	23:35.0	−2.75
3/15/2015	2161	170.48	−7.18	−24.36	32.168	23:38.7	−2.35
3/16/2015	2161	157.30	−7.16	−24.52	32.159	23:42.4	−1.96
3/17/2015	2161	144.11	−7.14	−24.67	32.150	23:46.0	−1.56
3/18/2015	2161	130.93	−7.11	−24.82	32.141	23:49.7	−1.17
3/19/2015	2161	117.74	−7.09	−24.96	32.132	23:53.3	−0.77
3/20/2015	2161	104.56	−7.06	−25.09	32.123	23:57.0	−0.37
3/21/2015	2161	91.37	−7.03	−25.21	32.114	00:0.7	0.02
3/22/2015	2161	78.18	−7.00	−25.33	32.105	00:4.3	0.42
3/23/2015	2161	65.00	−6.97	−25.44	32.096	00:7.9	0.81
3/24/2015	2161	51.81	−6.93	−25.55	32.087	00:11.6	1.21
3/25/2015	2161	38.68	−6.89	−25.65	32.078	00:15.2	1.60
3/26/2015	2161	25.49	−6.85	−25.74	32.069	00:18.9	1.99
3/27/2015	2161	12.30	−6.81	−25.82	32.060	00:22.5	2.39
3/28/2015	2162	359.11	−6.77	−25.90	32.050	00:26.2	2.78
3/29/2015	2162	345.92	−6.72	−25.97	32.041	00:29.8	3.17
3/30/2015	2162	332.73	−6.67	−26.03	32.032	00:33.4	3.56
3/31/2015	2162	319.53	−6.62	−26.09	32.023	00:37.1	3.95

(continued)

(continued)

Calendar Date	Rotation Number	Heliographic *Lo*	*Bo*	*Po*	Diameter (arcmin)	RA (HH:MM)	Dec (deg)
4/1/2015	2162	306.34	−6.57	−26.13	32.014	00:40.7	4.34
4/2/2015	2162	293.14	−6.52	−26.18	32.004	00:44.4	4.72
4/3/2015	2162	279.95	−6.46	−26.21	31.995	00:48.0	5.11
4/4/2015	2162	266.75	−6.40	−26.24	31.986	00:51.7	5.49
4/5/2015	2162	253.55	−6.34	−26.26	31.977	00:55.3	5.87
4/6/2015	2162	240.36	−6.28	−26.27	31.968	00:59.0	6.25
4/7/2015	2162	227.16	−6.22	−26.27	31.958	01:2.6	6.63
4/8/2015	2162	213.96	−6.15	−26.27	31.949	01:6.3	7.01
4/9/2015	2162	200.76	−6.09	−26.26	31.940	01:10.0	7.38
4/10/2015	2162	187.55	−6.02	−26.24	31.931	01:13.6	7.75
4/11/2015	2162	174.35	−5.95	−26.22	31.922	01:17.3	8.12
4/12/2015	2162	161.15	−5.87	−26.19	31.913	01:21.0	8.49
4/13/2015	2162	147.94	−5.80	−26.15	31.904	01:24.7	8.86
4/14/2015	2162	134.74	−5.73	−26.10	31.895	01:28.4	9.22
4/15/2015	2162	121.60	−5.65	−26.05	31.886	01:32.1	9.58
4/16/2015	2162	108.39	−5.57	−25.99	31.877	01:35.8	9.94
4/17/2015	2162	95.18	−5.49	−25.92	31.868	01:39.5	10.29
4/18/2015	2162	81.97	−5.41	−25.84	31.859	01:43.2	10.64
4/19/2015	2162	68.77	−5.32	−25.76	31.850	01:46.9	10.99
4/20/2015	2162	55.56	−5.24	−25.67	31.842	01:50.6	11.34
4/21/2015	2162	42.35	−5.15	−25.57	31.833	01:54.4	11.68
4/22/2015	2162	29.14	−5.07	−25.47	31.824	01:58.1	12.02
4/23/2015	2162	15.92	−4.98	−25.36	31.816	02:1.9	12.36
4/24/2015	2162	2.71	−4.89	−25.24	31.807	02:5.6	12.69
4/25/2015	2163	349.50	−4.80	−25.11	31.799	02:9.4	13.02
4/26/2015	2163	336.28	−4.70	−24.97	31.790	02:13.2	13.35
4/27/2015	2163	323.07	−4.61	−24.83	31.782	02:16.9	13.67
4/28/2015	2163	309.85	−4.51	−24.68	31.774	02:20.7	13.99
4/29/2015	2163	296.63	−4.41	−24.53	31.765	02:24.5	14.30
4/30/2015	2163	283.42	−4.32	−24.36	31.757	02:28.3	14.61
5/1/2015	2163	270.20	−4.22	−24.19	31.749	02:32.1	14.92
5/2/2015	2163	256.98	−4.12	−24.01	31.741	02:36.0	15.22
5/3/2015	2163	243.76	−4.02	−23.83	31.733	02:39.8	15.52
5/4/2015	2163	230.54	−3.91	−23.64	31.725	02:43.6	15.82
5/5/2015	2163	217.38	−3.81	−23.44	31.718	02:47.5	16.11
5/6/2015	2163	204.16	−3.70	−23.23	31.710	02:51.3	16.39
5/7/2015	2163	190.93	−3.60	−23.02	31.702	02:55.2	16.67
5/8/2015	2163	177.71	−3.49	−22.80	31.695	02:59.1	16.95

(continued)

(continued)

Calendar Date	Rotation Number	Heliographic Lo	Bo	Po	Diameter (arcmin)	RA (HH:MM)	Dec (deg)
5/9/2015	2163	164.49	−3.38	−22.57	31.687	03:3.0	17.22
5/10/2015	2163	151.26	−3.28	−22.33	31.680	03:6.9	17.49
5/11/2015	2163	138.04	−3.17	−22.09	31.673	03:10.8	17.75
5/12/2015	2163	124.81	−3.06	−21.84	31.666	03:14.7	18.01
5/13/2015	2163	111.58	−2.95	−21.59	31.659	03:18.6	18.26
5/14/2015	2163	98.35	−2.83	−21.33	31.652	03:22.5	18.50
5/15/2015	2163	85.13	−2.72	−21.06	31.645	03:26.5	18.75
5/16/2015	2163	71.90	−2.61	−20.78	31.638	03:30.4	18.98
5/17/2015	2163	58.67	−2.49	−20.50	31.632	03:34.4	19.21
5/18/2015	2163	45.44	−2.38	−20.21	31.625	03:38.4	19.44
5/19/2015	2163	32.21	−2.26	−19.92	31.619	03:42.3	19.66
5/20/2015	2163	18.98	−2.15	−19.62	31.612	03:46.3	19.87
5/21/2015	2163	5.75	−2.03	−19.31	31.606	03:50.3	20.08
5/22/2015	2164	352.51	−1.91	−19.00	31.600	03:54.3	20.28
5/23/2015	2164	339.28	−1.80	−18.68	31.594	03:58.4	20.48
5/24/2015	2164	326.05	−1.68	−18.36	31.588	04:2.4	20.67
5/25/2015	2164	312.88	−1.56	−18.03	31.583	04:6.4	20.86
5/26/2015	2164	299.64	−1.44	−17.69	31.577	04:10.5	21.04
5/27/2015	2164	286.41	−1.32	−17.35	31.571	04:14.5	21.21
5/28/2015	2164	273.17	−1.20	−17.00	31.566	04:18.6	21.38
5/29/2015	2164	259.94	−1.09	−16.65	31.561	04:22.6	21.54
5/30/2015	2164	246.70	−0.97	−16.29	31.556	04:26.7	21.69
5/31/2015	2164	233.46	−0.85	−15.93	31.551	04:30.8	21.84
6/1/2015	2164	220.23	−0.73	−15.56	31.546	04:34.9	21.99
6/2/2015	2164	206.99	−0.60	−15.19	31.541	04:39.0	22.12
6/3/2015	2164	193.75	−0.48	−14.81	31.536	04:43.1	22.25
6/4/2015	2164	180.52	−0.36	−14.42	31.532	04:47.2	22.37
6/5/2015	2164	167.28	−0.24	−14.04	31.528	04:51.3	22.49
6/6/2015	2164	154.04	−0.12	−13.65	31.523	04:55.4	22.60
6/7/2015	2164	140.80	−0.00	−13.25	31.519	04:59.5	22.70
6/8/2015	2164	127.56	0.12	−12.85	31.515	05:3.6	22.80
6/9/2015	2164	114.32	0.24	−12.45	31.512	05:7.8	22.89
6/10/2015	2164	101.08	0.36	−12.04	31.508	05:11.9	22.97
6/11/2015	2164	87.84	0.48	−11.63	31.504	05:16.0	23.05
6/12/2015	2164	74.60	0.60	−11.21	31.501	05:20.2	23.12
6/13/2015	2164	61.36	0.72	−10.79	31.498	05:24.3	23.18
6/14/2015	2164	48.19	0.84	−10.37	31.494	05:28.5	23.24
6/15/2015	2164	34.95	0.96	−9.95	31.491	05:32.6	23.29

(continued)

(continued)

Calendar Date	Rotation Number	Heliographic Lo	Bo	Po	Diameter (arcmin)	RA (HH:MM)	Dec (deg)
6/16/2015	2164	21.71	1.08	−9.52	31.489	05:36.8	23.33
6/17/2015	2164	8.47	1.20	−9.09	31.486	05:40.9	23.37
6/18/2015	2165	355.23	1.32	−8.66	31.483	05:45.1	23.39
6/19/2015	2165	341.99	1.44	−8.22	31.481	05:49.3	23.42
6/20/2015	2165	328.75	1.55	−7.78	31.479	05:53.4	23.43
6/21/2015	2165	315.50	1.67	−7.34	31.476	05:57.6	23.44
6/22/2015	2165	302.26	1.79	−6.90	31.474	06:1.7	23.44
6/23/2015	2165	289.02	1.90	−6.46	31.473	06:5.9	23.44
6/24/2015	2165	275.78	2.02	−6.01	31.471	06:10.0	23.42
6/25/2015	2165	262.54	2.14	−5.56	31.469	06:14.2	23.40
6/26/2015	2165	249.30	2.25	−5.12	31.468	06:18.3	23.38
6/27/2015	2165	236.06	2.36	−4.67	31.467	06:22.5	23.35
6/28/2015	2165	222.82	2.48	−4.21	31.466	06:26.6	23.31
6/29/2015	2165	209.58	2.59	−3.76	31.465	06:30.8	23.26
6/30/2015	2165	196.34	2.70	−3.31	31.464	06:34.9	23.21
7/1/2015	2165	183.10	2.81	−2.86	31.463	06:39.1	23.14
7/2/2015	2165	169.86	2.92	−2.41	31.463	06:43.2	23.08
7/3/2015	2165	156.62	3.03	−1.95	31.462	06:47.3	23.00
7/4/2015	2165	143.44	3.14	−1.50	31.462	06:51.5	22.92
7/5/2015	2165	130.20	3.25	−1.05	31.462	06:55.6	22.84
7/6/2015	2165	116.96	3.36	−0.59	31.462	06:59.7	22.74
7/7/2015	2165	103.72	3.46	−0.14	31.462	07:3.8	22.64
7/8/2015	2165	90.49	3.57	0.31	31.462	07:7.9	22.54
7/9/2015	2165	77.25	3.67	0.76	31.463	07:12.0	22.42
7/10/2015	2165	64.01	3.78	1.21	31.464	07:16.1	22.30
7/11/2015	2165	50.77	3.88	1.66	31.464	07:20.2	22.18
7/12/2015	2165	37.53	3.98	2.11	31.465	07:24.2	22.05
7/13/2015	2165	24.29	4.08	2.56	31.466	07:28.3	21.91
7/14/2015	2165	11.06	4.18	3.01	31.468	07:32.4	21.76
7/15/2015	2166	357.82	4.28	3.45	31.469	07:36.4	21.61
7/16/2015	2166	344.58	4.37	3.89	31.471	07:40.5	21.45
7/17/2015	2166	331.35	4.47	4.33	31.472	07:44.5	21.29
7/18/2015	2166	318.11	4.56	4.77	31.474	07:48.5	21.12
7/19/2015	2166	304.88	4.66	5.21	31.476	07:52.5	20.95
7/20/2015	2166	291.64	4.75	5.64	31.478	07:56.5	20.76
7/21/2015	2166	278.41	4.84	6.08	31.480	08:0.5	20.58
7/22/2015	2166	265.17	4.93	6.51	31.483	08:4.5	20.38
7/23/2015	2166	251.94	5.02	6.93	31.485	08:8.5	20.19

(continued)

(continued)

Calendar	Rotation	Heliographic			Diameter	RA	Dec
Date	Number	*Lo*	*Bo*	*Po*	(arcmin)	(HH:MM)	(deg)
7/24/2015	2166	238.77	5.10	7.36	31.488	08:12.5	19.98
7/25/2015	2166	225.53	5.19	7.78	31.491	08:16.4	19.77
7/26/2015	2166	212.30	5.27	8.20	31.494	08:20.4	19.56
7/27/2015	2166	199.07	5.35	8.62	31.497	08:24.3	19.34
7/28/2015	2166	185.84	5.43	9.03	31.500	08:28.3	19.11
7/29/2015	2166	172.61	5.51	9.44	31.504	08:32.2	18.88
7/30/2015	2166	159.38	5.59	9.85	31.507	08:36.1	18.64
7/31/2015	2166	146.15	5.67	10.25	31.511	08:40.0	18.40
8/1/2015	2166	132.92	5.74	10.65	31.515	08:43.9	18.16
8/2/2015	2166	119.69	5.82	11.05	31.519	08:47.8	17.90
8/3/2015	2166	106.46	5.89	11.44	31.523	08:51.6	17.65
8/4/2015	2166	93.23	5.96	11.83	31.527	08:55.5	17.39
8/5/2015	2166	80.00	6.03	12.22	31.531	08:59.4	17.12
8/6/2015	2166	66.78	6.09	12.60	31.536	09:3.2	16.85
8/7/2015	2166	53.55	6.16	12.98	31.540	09:7.0	16.58
8/8/2015	2166	40.32	6.22	13.35	31.545	09:10.9	16.30
8/9/2015	2166	27.10	6.28	13.72	31.550	09:14.7	16.01
8/10/2015	2166	13.87	6.34	14.09	31.555	09:18.5	15.73
8/11/2015	2166	0.65	6.40	14.45	31.560	09:22.3	15.43
8/12/2015	2167	347.43	6.46	14.81	31.565	09:26.0	15.14
8/13/2015	2167	334.26	6.51	15.16	31.570	09:29.8	14.84
8/14/2015	2167	321.04	6.56	15.51	31.576	09:33.6	14.53
8/15/2015	2167	307.82	6.61	15.85	31.582	09:37.3	14.23
8/16/2015	2167	294.60	6.66	16.19	31.587	09:41.1	13.91
8/17/2015	2167	281.38	6.71	16.53	31.593	09:44.8	13.60
8/18/2015	2167	268.16	6.75	16.86	31.599	09:48.6	13.28
8/19/2015	2167	254.94	6.80	17.18	31.605	09:52.3	12.96
8/20/2015	2167	241.72	6.84	17.50	31.611	09:56.0	12.63
8/21/2015	2167	228.50	6.88	17.82	31.618	09:59.7	12.30
8/22/2015	2167	215.28	6.92	18.13	31.624	10:3.4	11.97
8/23/2015	2167	202.06	6.95	18.43	31.630	10:7.1	11.63
8/24/2015	2167	188.85	6.99	18.73	31.637	10:10.8	11.29
8/25/2015	2167	175.63	7.02	19.03	31.644	10:14.4	10.95
8/26/2015	2167	162.41	7.05	19.32	31.651	10:18.1	10.61
8/27/2015	2167	149.20	7.08	19.61	31.657	10:21.8	10.26
8/28/2015	2167	135.98	7.10	19.89	31.664	10:25.4	9.91
8/29/2015	2167	122.77	7.12	20.16	31.672	10:29.1	9.55
8/30/2015	2167	109.56	7.15	20.43	31.679	10:32.7	9.20

(continued)

(continued)

Calendar	Rotation	Heliographic			Diameter	RA	Dec
Date	Number	*Lo*	*Bo*	*Po*	(arcmin)	(HH:MM)	(deg)
8/31/2015	2167	96.34	7.17	20.69	31.686	10:36.4	8.84
9/1/2015	2167	83.13	7.18	20.95	31.694	10:40.0	8.48
9/2/2015	2167	69.92	7.20	21.21	31.701	10:43.6	8.12
9/3/2015	2167	56.77	7.21	21.45	31.709	10:47.2	7.75
9/4/2015	2167	43.56	7.22	21.70	31.716	10:50.9	7.39
9/5/2015	2167	30.35	7.23	21.93	31.724	10:54.5	7.02
9/6/2015	2167	17.14	7.24	22.16	31.732	10:58.1	6.65
9/7/2015	2167	3.93	7.25	22.39	31.740	11:1.7	6.27
9/8/2015	2168	350.72	7.25	22.61	31.748	11:5.3	5.90
9/9/2015	2168	337.51	7.25	22.82	31.756	11:8.9	5.52
9/10/2015	2168	324.30	7.25	23.03	31.764	11:12.5	5.15
9/11/2015	2168	311.10	7.25	23.23	31.772	11:16.1	4.77
9/12/2015	2168	297.89	7.24	23.42	31.780	11:19.7	4.39
9/13/2015	2168	284.68	7.23	23.61	31.789	11:23.3	4.01
9/14/2015	2168	271.48	7.22	23.80	31.797	11:26.9	3.62
9/15/2015	2168	258.27	7.21	23.97	31.806	11:30.5	3.24
9/16/2015	2168	245.07	7.20	24.14	31.814	11:34.0	2.85
9/17/2015	2168	231.86	7.18	24.31	31.823	11:37.6	2.47
9/18/2015	2168	218.66	7.16	24.47	31.831	11:41.2	2.08
9/19/2015	2168	205.46	7.14	24.62	31.840	11:44.8	1.69
9/20/2015	2168	192.25	7.12	24.76	31.849	11:48.4	1.31
9/21/2015	2168	179.05	7.10	24.90	31.858	11:52.0	0.92
9/22/2015	2168	165.85	7.07	25.04	31.866	11:55.6	0.53
9/23/2015	2168	152.71	7.04	25.16	31.875	11:59.2	0.14
9/24/2015	2168	139.51	7.01	25.28	31.884	12:2.8	−0.25
9/25/2015	2168	126.31	6.98	25.40	31.893	12:6.4	−0.64
9/26/2015	2168	113.11	6.95	25.50	31.902	12:10.0	−1.03
9/27/2015	2168	99.91	6.91	25.60	31.911	12:13.6	−1.42
9/28/2015	2168	86.71	6.87	25.70	31.920	12:17.2	−1.81
9/29/2015	2168	73.51	6.83	25.78	31.929	12:20.8	−2.20
9/30/2015	2168	60.31	6.79	25.86	31.938	12:24.4	−2.59
10/1/2015	2168	47.11	6.74	25.93	31.948	12:28.0	−2.98
10/2/2015	2168	33.91	6.70	26.00	31.957	12:31.6	−3.37
10/3/2015	2168	20.72	6.65	26.06	31.966	12:35.3	−3.75
10/4/2015	2168	7.52	6.60	26.11	31.975	12:38.9	−4.14
10/5/2015	2169	354.32	6.54	26.15	31.984	12:42.5	−4.53
10/6/2015	2169	341.13	6.49	26.19	31.993	12:46.2	−4.91
10/7/2015	2169	327.93	6.43	26.22	32.003	12:49.8	−5.30

(continued)

(continued)

Calendar	Rotation	Heliographic			Diameter	RA	Dec
Date	Number	*Lo*	*Bo*	*Po*	(arcmin)	(HH:MM)	(deg)
10/8/2015	2169	314.74	6.37	26.25	32.012	12:53.5	−5.68
10/9/2015	2169	301.54	6.31	26.26	32.021	12:57.2	−6.06
10/10/2015	2169	288.34	6.25	26.27	32.030	13:0.8	−6.44
10/11/2015	2169	275.15	6.18	26.27	32.040	13:4.5	−6.82
10/12/2015	2169	261.96	6.12	26.27	32.049	13:8.2	−7.20
10/13/2015	2169	248.82	6.05	26.25	32.058	13:11.9	−7.57
10/14/2015	2169	235.63	5.98	26.23	32.067	13:15.6	−7.95
10/15/2015	2169	222.44	5.91	26.20	32.076	13:19.3	−8.32
10/16/2015	2169	209.24	5.83	26.17	32.085	13:23.0	−8.69
10/17/2015	2169	196.05	5.76	26.12	32.095	13:26.8	−9.06
10/18/2015	2169	182.86	5.68	26.07	32.104	13:30.5	−9.43
10/19/2015	2169	169.67	5.60	26.01	32.113	13:34.2	−9.79
10/20/2015	2169	156.47	5.52	25.95	32.122	13:38.0	−10.15
10/21/2015	2169	143.28	5.44	25.87	32.131	13:41.8	−10.51
10/22/2015	2169	130.09	5.36	25.79	32.140	13:45.6	−10.87
10/23/2015	2169	116.90	5.27	25.70	32.149	13:49.4	−11.22
10/24/2015	2169	103.71	5.18	25.61	32.158	13:53.2	−11.57
10/25/2015	2169	90.52	5.09	25.50	32.166	13:57.0	−11.92
10/26/2015	2169	77.33	5.00	25.39	32.175	14:0.8	−12.26
10/27/2015	2169	64.14	4.91	25.27	32.184	14:4.6	−12.61
10/28/2015	2169	50.95	4.82	25.14	32.193	14:8.5	−12.94
10/29/2015	2169	37.76	4.72	25.00	32.201	14:12.4	−13.28
10/30/2015	2169	24.57	4.63	24.86	32.210	14:16.2	−13.61
10/31/2015	2169	11.38	4.53	24.71	32.218	14:20.1	−13.94
11/1/2015	2170	358.19	4.43	24.55	32.227	14:24.0	−14.26
11/2/2015	2170	345.06	4.33	24.38	32.235	14:27.9	−14.58
11/3/2015	2170	331.88	4.22	24.20	32.244	14:31.9	−14.90
11/4/2015	2170	318.69	4.12	24.02	32.252	14:35.8	−15.21
11/5/2015	2170	305.50	4.02	23.83	32.260	14:39.8	−15.52
11/6/2015	2170	292.31	3.91	23.63	32.268	14:43.7	−15.83
11/7/2015	2170	279.12	3.80	23.42	32.276	14:47.7	−16.13
11/8/2015	2170	265.94	3.69	23.21	32.284	14:51.7	−16.42
11/9/2015	2170	252.75	3.58	22.99	32.292	14:55.7	−16.71
11/10/2015	2170	239.56	3.47	22.76	32.300	14:59.8	−17.00
11/11/2015	2170	226.37	3.36	22.52	32.307	15:3.8	−17.28
11/12/2015	2170	213.19	3.25	22.27	32.315	15:7.9	−17.56
11/13/2015	2170	200.00	3.13	22.02	32.322	15:11.9	−17.83
11/14/2015	2170	186.82	3.02	21.76	32.330	15:16.0	−18.09

(continued)

(continued)

Calendar	Rotation	Heliographic			Diameter	RA	Dec
Date	Number	*Lo*	*Bo*	*Po*	(arcmin)	(HH:MM)	(deg)
11/15/2015	2170	173.63	2.90	21.49	32.337	15:20.1	−18.35
11/16/2015	2170	160.44	2.79	21.21	32.344	15:24.2	−18.61
11/17/2015	2170	147.26	2.67	20.93	32.351	15:28.3	−18.86
11/18/2015	2170	134.07	2.55	20.64	32.358	15:32.5	−19.10
11/19/2015	2170	120.89	2.43	20.34	32.365	15:36.6	−19.34
11/20/2015	2170	107.70	2.31	20.03	32.372	15:40.8	−19.57
11/21/2015	2170	94.52	2.19	19.72	32.379	15:45.0	−19.80
11/22/2015	2170	81.40	2.07	19.40	32.385	15:49.2	−20.02
11/23/2015	2170	68.21	1.94	19.08	32.392	15:53.4	−20.24
11/24/2015	2170	55.03	1.82	18.74	32.398	15:57.6	−20.44
11/25/2015	2170	41.85	1.70	18.40	32.404	16:1.8	−20.65
11/26/2015	2170	28.66	1.57	18.05	32.410	16:6.1	−20.84
11/27/2015	2170	15.48	1.45	17.70	32.416	16:10.3	−21.03
11/28/2015	2170	2.30	1.32	17.34	32.422	16:14.6	−21.21
11/29/2015	2171	349.11	1.20	16.97	32.428	16:18.9	−21.39
11/30/2015	2171	335.93	1.07	16.60	32.433	16:23.2	−21.56
12/1/2015	2171	322.75	0.94	16.22	32.439	16:27.5	−21.72
12/2/2015	2171	309.57	0.82	15.84	32.444	16:31.8	−21.88
12/3/2015	2171	296.39	0.69	15.45	32.449	16:36.1	−22.03
12/4/2015	2171	283.20	0.56	15.05	32.454	16:40.4	−22.17
12/5/2015	2171	270.02	0.43	14.65	32.459	16:44.8	−22.30
12/6/2015	2171	256.84	0.31	14.24	32.463	16:49.1	−22.43
12/7/2015	2171	243.66	0.18	13.83	32.468	16:53.5	−22.55
12/8/2015	2171	230.48	0.05	13.41	32.472	16:57.9	−22.66
12/9/2015	2171	217.30	−0.08	12.98	32.477	17:2.3	−22.77
12/10/2015	2171	204.12	−0.21	12.56	32.481	17:6.6	−22.87
12/11/2015	2171	190.94	−0.33	12.13	32.485	17:11.0	−22.96
12/12/2015	2171	177.82	−0.46	11.69	32.488	17:15.4	−23.04
12/13/2015	2171	164.64	−0.59	11.25	32.492	17:19.8	−23.11
12/14/2015	2171	151.46	−0.72	10.80	32.496	17:24.3	−23.18
12/15/2015	2171	138.28	−0.85	10.35	32.499	17:28.7	−23.24
12/16/2015	2171	125.11	−0.97	9.90	32.502	17:33.1	−23.29
12/17/2015	2171	111.93	−1.10	9.44	32.505	17:37.5	−23.34
12/18/2015	2171	98.75	−1.23	8.98	32.508	17:41.9	−23.37
12/19/2015	2171	85.57	−1.35	8.52	32.511	17:46.4	−23.40
12/20/2015	2171	72.39	−1.48	8.06	32.513	17:50.8	−23.42
12/21/2015	2171	59.22	−1.60	7.59	32.516	17:55.2	−23.44
12/22/2015	2171	46.04	−1.73	7.12	32.518	17:59.7	−23.44

(continued)

(continued)

Calendar	Rotation	Heliographic			Diameter	RA	Dec
Date	Number	*Lo*	*Bo*	*Po*	(arcmin)	(HH:MM)	(deg)
12/23/2015	2171	32.86	−1.85	6.65	32.520	18:4.1	−23.44
12/24/2015	2171	19.69	−1.98	6.17	32.522	18:8.6	−23.43
12/25/2015	2171	6.51	−2.10	5.69	32.523	18:13.0	−23.41
12/26/2015	2172	353.33	−2.22	5.21	32.525	18:17.4	−23.38
12/27/2015	2172	340.16	−2.35	4.73	32.526	18:21.9	−23.35
12/28/2015	2172	326.98	−2.47	4.25	32.528	18:26.3	−23.31
12/29/2015	2172	313.81	−2.59	3.77	32.529	18:30.7	−23.26
12/30/2015	2172	300.63	−2.71	3.29	32.530	18:35.2	−23.20
12/31/2015	2172	287.46	−2.83	2.80	32.530	18:39.6	−23.14
1/1/2016	2172	274.28	−2.94	2.32	32.531	18:44.0	−23.06
1/2/2016	2172	261.17	−3.06	1.83	32.531	18:48.4	−22.98
1/3/2016	2172	248.00	−3.18	1.35	32.531	18:52.8	−22.90
1/4/2016	2172	234.82	−3.29	0.87	32.531	18:57.2	−22.80
1/5/2016	2172	221.65	−3.41	0.38	32.531	19:1.6	−22.70
1/6/2016	2172	208.48	−3.52	−0.10	32.531	19:6.0	−22.59
1/7/2016	2172	195.30	−3.63	−0.59	32.530	19:10.4	−22.47
1/8/2016	2172	182.13	−3.74	−1.07	32.530	19:14.8	−22.34
1/9/2016	2172	168.96	−3.85	−1.55	32.529	19:19.1	−22.21
1/10/2016	2172	155.78	−3.96	−2.03	32.528	19:23.5	−22.07
1/11/2016	2172	142.61	−4.07	−2.51	32.527	19:27.8	−21.92
1/12/2016	2172	129.44	−4.17	−2.98	32.526	19:32.2	−21.77
1/13/2016	2172	116.27	−4.28	−3.46	32.524	19:36.5	−21.61
1/14/2016	2172	103.10	−4.38	−3.93	32.522	19:40.8	−21.44
1/15/2016	2172	89.93	−4.48	−4.40	32.521	19:45.1	−21.26
1/16/2016	2172	76.75	−4.58	−4.87	32.519	19:49.4	−21.08
1/17/2016	2172	63.58	−4.68	−5.33	32.517	19:53.7	−20.89
1/18/2016	2172	50.41	−4.78	−5.80	32.514	19:58.0	−20.70
1/19/2016	2172	37.24	−4.88	−6.26	32.512	20:2.2	−20.50
1/20/2016	2172	24.07	−4.97	−6.72	32.509	20:6.5	−20.29
1/21/2016	2172	10.90	−5.06	−7.17	32.506	20:10.7	−20.07
1/22/2016	2173	357.79	−5.16	−7.62	32.503	20:14.9	−19.85
1/23/2016	2173	344.62	−5.24	−8.07	32.500	20:19.2	−19.63
1/24/2016	2173	331.45	−5.33	−8.51	32.497	20:23.4	−19.39
1/25/2016	2173	318.28	−5.42	−8.95	32.494	20:27.5	−19.15
1/26/2016	2173	305.11	−5.50	−9.39	32.490	20:31.7	−18.91
1/27/2016	2173	291.94	−5.59	−9.82	32.486	20:35.9	−18.66
1/28/2016	2173	278.77	−5.67	−10.25	32.482	20:40.0	−18.40
1/29/2016	2173	265.60	−5.75	−10.68	32.478	20:44.2	−18.14

(continued)

(continued)

Calendar	Rotation	Heliographic			Diameter	RA	Dec
Date	Number	Lo	Bo	Po	(arcmin)	(HH:MM)	(deg)
1/30/2016	2173	252.43	−5.82	−11.10	32.474	20:48.3	−17.87
1/31/2016	2173	239.26	−5.90	−11.52	32.470	20:52.4	−17.60
2/1/2016	2173	226.09	−5.97	−11.93	32.465	20:56.5	−17.32
2/2/2016	2173	212.92	−6.05	−12.34	32.461	21:0.6	−17.04
2/3/2016	2173	199.75	−6.12	−12.74	32.456	21:4.6	−16.75
2/4/2016	2173	186.58	−6.18	−13.14	32.451	21:8.7	−16.46
2/5/2016	2173	173.41	−6.25	−13.53	32.446	21:12.7	−16.16
2/6/2016	2173	160.24	−6.31	−13.92	32.441	21:16.7	−15.86
2/7/2016	2173	147.07	−6.38	−14.30	32.435	21:20.8	−15.55
2/8/2016	2173	133.90	−6.44	−14.68	32.430	21:24.7	−15.24
2/9/2016	2173	120.73	−6.49	−15.06	32.424	21:28.7	−14.92
2/10/2016	2173	107.56	−6.55	−15.43	32.419	21:32.7	−14.60
2/11/2016	2173	94.45	−6.60	−15.79	32.413	21:36.7	−14.28
2/12/2016	2173	81.28	−6.66	−16.15	32.407	21:40.6	−13.95
2/13/2016	2173	68.11	−6.71	−16.50	32.401	21:44.6	−13.62
2/14/2016	2173	54.94	−6.75	−16.85	32.394	21:48.5	−13.29
2/15/2016	2173	41.77	−6.80	−17.19	32.388	21:52.4	−12.95
2/16/2016	2173	28.60	−6.84	−17.53	32.382	21:56.3	−12.60
2/17/2016	2173	15.43	−6.88	−17.86	32.375	21:60.2	−12.26
2/18/2016	2173	2.25	−6.92	−18.18	32.368	22:4.0	−11.91
2/19/2016	2174	349.08	−6.96	−18.50	32.361	22:7.9	−11.56
2/20/2016	2174	335.91	−6.99	−18.81	32.354	22:11.8	−11.20
2/21/2016	2174	322.74	−7.03	−19.12	32.347	22:15.6	−10.84
2/22/2016	2174	309.57	−7.06	−19.42	32.340	22:19.4	−10.48
2/23/2016	2174	296.39	−7.09	−19.72	32.333	22:23.2	−10.12
2/24/2016	2174	283.22	−7.11	−20.01	32.326	22:27.0	−9.75
2/25/2016	2174	270.05	−7.14	−20.29	32.318	22:30.8	−9.38
2/26/2016	2174	256.87	−7.16	−20.57	32.311	22:34.6	−9.01
2/27/2016	2174	243.70	−7.18	−20.84	32.303	22:38.4	−8.64
2/28/2016	2174	230.52	−7.19	−21.11	32.295	22:42.2	−8.26
2/29/2016	2174	217.35	−7.21	−21.36	32.287	22:45.9	−7.89
3/1/2016	2174	204.17	−7.22	−21.62	32.279	22:49.7	−7.51
3/2/2016	2174	191.06	−7.23	−21.86	32.272	22:53.4	−7.12
3/3/2016	2174	177.88	−7.24	−22.10	32.263	22:57.2	−6.74
3/4/2016	2174	164.71	−7.24	−22.34	32.255	23:0.9	−6.36
3/5/2016	2174	151.53	−7.25	−22.56	32.247	23:4.6	−5.97
3/6/2016	2174	138.35	−7.25	−22.79	32.239	23:8.3	−5.58
3/7/2016	2174	125.17	−7.25	−23.00	32.230	23:12.0	−5.19

(continued)

(continued)

Calendar	Rotation	Heliographic			Diameter	RA	Dec
Date	Number	*Lo*	*Bo*	*Po*	(arcmin)	(HH:MM)	(deg)
3/8/2016	2174	111.99	−7.25	−23.21	32.222	23:15.7	−4.80
3/9/2016	2174	98.81	−7.24	−23.41	32.213	23:19.4	−4.41
3/10/2016	2174	85.63	−7.23	−23.60	32.205	23:23.1	−4.02
3/11/2016	2174	72.45	−7.22	−23.79	32.196	23:26.8	−3.63
3/12/2016	2174	59.27	−7.21	−23.97	32.188	23:30.5	−3.24
3/13/2016	2174	46.09	−7.20	−24.15	32.179	23:34.2	−2.84
3/14/2016	2174	32.91	−7.18	−24.32	32.170	23:37.8	−2.45
3/15/2016	2174	19.73	−7.16	−24.48	32.161	23:41.5	−2.05
3/16/2016	2174	6.54	−7.14	−24.63	32.152	23:45.2	−1.66
3/17/2016	2175	353.36	−7.12	−24.78	32.143	23:48.8	−1.26
3/18/2016	2175	340.17	−7.10	−24.92	32.135	23:52.5	−0.87
3/19/2016	2175	326.99	−7.07	−25.06	32.126	23:56.1	−0.47
3/20/2016	2175	313.80	−7.04	−25.18	32.116	23:59.8	−0.07
3/21/2016	2175	300.62	−7.01	−25.30	32.107	00:3.4	0.32
3/22/2016	2175	287.49	−6.97	−25.42	32.098	00:7.1	0.72
3/23/2016	2175	274.30	−6.94	−25.52	32.089	00:10.7	1.11
3/24/2016	2175	261.11	−6.90	−25.62	32.080	00:14.3	1.51
3/25/2016	2175	247.92	−6.86	−25.71	32.071	00:18.0	1.90
3/26/2016	2175	234.73	−6.82	−25.80	32.062	00:21.6	2.29
3/27/2016	2175	221.54	−6.78	−25.88	32.053	00:25.3	2.68
3/28/2016	2175	208.35	−6.73	−25.95	32.043	00:28.9	3.07
3/29/2016	2175	195.16	−6.68	−26.01	32.034	00:32.6	3.46
3/30/2016	2175	181.97	−6.63	−26.07	32.025	00:36.2	3.85
3/31/2016	2175	168.77	−6.58	−26.12	32.016	00:39.8	4.24
4/1/2016	2175	155.58	−6.53	−26.16	32.007	00:43.5	4.63
4/2/2016	2175	142.38	−6.47	−26.20	31.997	00:47.1	5.01
4/3/2016	2175	129.18	−6.42	−26.23	31.988	00:50.8	5.40
4/4/2016	2175	115.99	−6.36	−26.25	31.979	00:54.4	5.78
4/5/2016	2175	102.79	−6.30	−26.26	31.970	00:58.1	6.16
4/6/2016	2175	89.59	−6.23	−26.27	31.961	01:1.8	6.54
4/7/2016	2175	76.39	−6.17	−26.27	31.951	01:5.4	6.92
4/8/2016	2175	63.19	−6.10	−26.26	31.942	01:9.1	7.29
4/9/2016	2175	49.99	−6.03	−26.25	31.933	01:12.8	7.66
4/10/2016	2175	36.79	−5.96	−26.22	31.924	01:16.4	8.03
4/11/2016	2175	23.65	−5.89	−26.19	31.915	01:20.1	8.40
4/12/2016	2175	10.44	−5.82	−26.16	31.906	01:23.8	8.77
4/13/2016	2176	357.24	−5.74	−26.11	31.897	01:27.5	9.13
4/14/2016	2176	344.03	−5.67	−26.06	31.888	01:31.2	9.49

(continued)

(continued)

Calendar	Rotation	Heliographic			Diameter	RA	Dec
Date	Number	*Lo*	*Bo*	*Po*	(arcmin)	(HH:MM)	(deg)
4/15/2016	2176	330.83	−5.59	−26.00	31.879	01:34.9	9.85
4/16/2016	2176	317.62	−5.51	−25.94	31.870	01:38.6	10.21
4/17/2016	2176	304.41	−5.43	−25.86	31.861	01:42.3	10.56
4/18/2016	2176	291.20	−5.35	−25.78	31.852	01:46.0	10.91
4/19/2016	2176	277.99	−5.26	−25.69	31.844	01:49.7	11.26
4/20/2016	2176	264.78	−5.18	−25.60	31.835	01:53.5	11.60
4/21/2016	2176	251.57	−5.09	−25.49	31.826	01:57.2	11.94
4/22/2016	2176	238.36	−5.00	−25.38	31.818	02:1.0	12.28
4/23/2016	2176	225.15	−4.91	−25.26	31.809	02:4.7	12.61
4/24/2016	2176	211.93	−4.82	−25.14	31.801	02:8.5	12.94
4/25/2016	2176	198.72	−4.73	−25.01	31.792	02:12.2	13.27
4/26/2016	2176	185.50	−4.63	−24.87	31.784	02:16.0	13.59
4/27/2016	2176	172.29	−4.54	−24.72	31.776	02:19.8	13.91
4/28/2016	2176	159.07	−4.44	−24.56	31.767	02:23.6	14.23
4/29/2016	2176	145.85	−4.34	−24.40	31.759	02:27.4	14.54
4/30/2016	2176	132.64	−4.24	−24.23	31.751	02:31.2	14.85
5/1/2016	2176	119.48	−4.14	−24.06	31.743	02:35.0	15.15
5/2/2016	2176	106.26	−4.04	−23.87	31.735	02:38.9	15.45
5/3/2016	2176	93.04	−3.94	−23.68	31.727	02:42.7	15.75
5/4/2016	2176	79.82	−3.84	−23.48	31.719	02:46.5	16.04
5/5/2016	2176	66.60	−3.73	−23.28	31.712	02:50.4	16.32
5/6/2016	2176	53.37	−3.63	−23.07	31.704	02:54.3	16.61
5/7/2016	2176	40.15	−3.52	−22.85	31.697	02:58.1	16.88
5/8/2016	2176	26.93	−3.41	−22.62	31.689	03:2.0	17.16
5/9/2016	2176	13.70	−3.30	−22.39	31.682	03:5.9	17.42
5/10/2016	2177	0.48	−3.19	−22.15	31.675	03:9.8	17.69
5/11/2016	2177	347.25	−3.08	−21.90	31.667	03:13.7	17.94
5/12/2016	2177	334.02	−2.97	−21.65	31.660	03:17.7	18.20
5/13/2016	2177	320.80	−2.86	−21.39	31.653	03:21.6	18.45
5/14/2016	2177	307.57	−2.75	−21.12	31.647	03:25.5	18.69
5/15/2016	2177	294.34	−2.64	−20.85	31.640	03:29.5	18.93
5/16/2016	2177	281.11	−2.52	−20.57	31.633	03:33.4	19.16
5/17/2016	2177	267.88	−2.41	−20.28	31.627	03:37.4	19.38
5/18/2016	2177	254.65	−2.29	−19.99	31.620	03:41.4	19.61
5/19/2016	2177	241.42	−2.18	−19.69	31.614	03:45.4	19.82
5/20/2016	2177	228.19	−2.06	−19.39	31.608	03:49.4	20.03
5/21/2016	2177	214.96	−1.94	−19.08	31.602	03:53.4	20.24
5/22/2016	2177	201.79	−1.83	−18.76	31.596	03:57.4	20.43

(continued)

(continued)

Calendar	Rotation	Heliographic			Diameter	RA	Dec
Date	Number	*Lo*	*Bo*	*Po*	(arcmin)	(HH:MM)	(deg)
5/23/2016	2177	188.55	−1.71	−18.44	31.590	04:1.4	20.63
5/24/2016	2177	175.32	−1.59	−18.11	31.584	04:5.4	20.81
5/25/2016	2177	162.08	−1.47	−17.77	31.578	04:9.5	20.99
5/26/2016	2177	148.85	−1.35	−17.43	31.573	04:13.5	21.17
5/27/2016	2177	135.62	−1.24	−17.09	31.567	04:17.6	21.34
5/28/2016	2177	122.38	−1.12	−16.73	31.562	04:21.6	21.50
5/29/2016	2177	109.14	−1.00	−16.38	31.557	04:25.7	21.66
5/30/2016	2177	95.91	−0.88	−16.02	31.552	04:29.8	21.81
5/31/2016	2177	82.67	−0.76	−15.65	31.547	04:33.9	21.95
6/1/2016	2177	69.44	−0.64	−15.28	31.542	04:38.0	22.09
6/2/2016	2177	56.20	−0.52	−14.90	31.538	04:42.1	22.22
6/3/2016	2177	42.96	−0.39	−14.52	31.533	04:46.2	22.34
6/4/2016	2177	29.72	−0.27	−14.13	31.529	04:50.3	22.46
6/5/2016	2177	16.48	−0.15	−13.74	31.524	04:54.4	22.57
6/6/2016	2177	3.25	−0.03	−13.35	31.520	04:58.5	22.68
6/7/2016	2178	350.01	0.09	−12.95	31.516	05:2.6	22.78
6/8/2016	2178	336.77	0.21	−12.54	31.512	05:6.8	22.87
6/9/2016	2178	323.53	0.33	−12.14	31.509	05:10.9	22.95
6/10/2016	2178	310.29	0.45	−11.73	31.505	05:15.0	23.03
6/11/2016	2178	297.11	0.57	−11.31	31.502	05:19.2	23.10
6/12/2016	2178	283.87	0.69	−10.90	31.498	05:23.3	23.17
6/13/2016	2178	270.63	0.81	−10.47	31.495	05:27.5	23.23
6/14/2016	2178	257.39	0.93	−10.05	31.492	05:31.6	23.28
6/15/2016	2178	244.15	1.05	−9.62	31.489	05:35.8	23.32
6/16/2016	2178	230.91	1.17	−9.19	31.487	05:39.9	23.36
6/17/2016	2178	217.67	1.29	−8.76	31.484	05:44.1	23.39
6/18/2016	2178	204.43	1.41	−8.33	31.481	05:48.3	23.41
6/19/2016	2178	191.19	1.52	−7.89	31.479	05:52.4	23.43
6/20/2016	2178	177.95	1.64	−7.45	31.477	05:56.6	23.44
6/21/2016	2178	164.71	1.76	−7.01	31.475	06:0.7	23.44
6/22/2016	2178	151.47	1.87	−6.56	31.473	06:4.9	23.44
6/23/2016	2178	138.23	1.99	−6.12	31.471	06:9.0	23.43
6/24/2016	2178	124.99	2.11	−5.67	31.470	06:13.2	23.41
6/25/2016	2178	111.75	2.22	−5.22	31.468	06:17.3	23.39
6/26/2016	2178	98.50	2.33	−4.78	31.467	06:21.5	23.35
6/27/2016	2178	85.26	2.45	−4.32	31.466	06:25.6	23.32
6/28/2016	2178	72.02	2.56	−3.87	31.465	06:29.8	23.27
6/29/2016	2178	58.78	2.67	−3.42	31.464	06:33.9	23.22

(continued)

(continued)

Calendar	Rotation	Heliographic			Diameter	RA	Dec
Date	Number	*Lo*	*Bo*	*Po*	(arcmin)	(HH:MM)	(deg)
6/30/2016	2178	45.54	2.78	−2.97	31.463	06:38.1	23.16
7/1/2016	2178	32.37	2.90	−2.52	31.463	06:42.2	23.09
7/2/2016	2178	19.13	3.01	−2.06	31.462	06:46.3	23.02
7/3/2016	2178	5.89	3.11	−1.61	31.462	06:50.5	22.94
7/4/2016	2179	352.65	3.22	−1.16	31.462	06:54.6	22.86
7/5/2016	2179	339.41	3.33	−0.70	31.462	06:58.7	22.77
7/6/2016	2179	326.17	3.44	−0.25	31.462	07:2.8	22.67
7/7/2016	2179	312.93	3.54	0.20	31.462	07:6.9	22.56
7/8/2016	2179	299.69	3.65	0.65	31.463	07:11.0	22.45
7/9/2016	2179	286.45	3.75	1.10	31.463	07:15.1	22.33
7/10/2016	2179	273.21	3.85	1.55	31.464	07:19.2	22.21
7/11/2016	2179	259.98	3.95	2.00	31.465	07:23.3	22.08
7/12/2016	2179	246.74	4.05	2.45	31.466	07:27.3	21.94
7/13/2016	2179	233.50	4.15	2.90	31.467	07:31.4	21.80
7/14/2016	2179	220.26	4.25	3.34	31.469	07:35.4	21.65
7/15/2016	2179	207.03	4.35	3.78	31.470	07:39.5	21.49
7/16/2016	2179	193.79	4.45	4.23	31.472	07:43.5	21.33
7/17/2016	2179	180.56	4.54	4.67	31.474	07:47.5	21.16
7/18/2016	2179	167.32	4.63	5.10	31.476	07:51.6	20.99
7/19/2016	2179	154.08	4.73	5.54	31.478	07:55.6	20.81
7/20/2016	2179	140.85	4.82	5.97	31.480	07:59.6	20.62
7/21/2016	2179	127.68	4.91	6.40	31.482	08:3.6	20.43
7/22/2016	2179	114.44	4.99	6.83	31.485	08:7.5	20.23
7/23/2016	2179	101.21	5.08	7.26	31.487	08:11.5	20.03
7/24/2016	2179	87.98	5.17	7.68	31.490	08:15.5	19.82
7/25/2016	2179	74.74	5.25	8.10	31.493	08:19.4	19.61
7/26/2016	2179	61.51	5.33	8.52	31.496	08:23.4	19.39
7/27/2016	2179	48.28	5.41	8.93	31.499	08:27.3	19.17
7/28/2016	2179	35.05	5.49	9.34	31.503	08:31.2	18.94
7/29/2016	2179	21.82	5.57	9.75	31.506	08:35.2	18.70
7/30/2016	2179	8.59	5.65	10.15	31.510	08:39.1	18.46
7/31/2016	2180	355.36	5.72	10.55	31.514	08:43.0	18.22
8/1/2016	2180	342.13	5.80	10.95	31.518	08:46.8	17.97
8/2/2016	2180	328.90	5.87	11.35	31.522	08:50.7	17.71
8/3/2016	2180	315.67	5.94	11.74	31.526	08:54.6	17.45
8/4/2016	2180	302.44	6.01	12.12	31.530	08:58.4	17.19
8/5/2016	2180	289.22	6.07	12.51	31.535	09:2.3	16.92
8/6/2016	2180	275.99	6.14	12.89	31.539	09:6.1	16.64

(continued)

(continued)

Calendar	Rotation	Heliographic			Diameter	RA	Dec
Date	Number	*Lo*	*Bo*	*Po*	(arcmin)	(HH:MM)	(deg)
8/7/2016	2180	262.76	6.20	13.26	31.544	09:9.9	16.37
8/8/2016	2180	249.54	6.27	13.63	31.549	09:13.7	16.08
8/9/2016	2180	236.31	6.33	14.00	31.554	09:17.6	15.80
8/10/2016	2180	223.15	6.38	14.36	31.559	09:21.3	15.51
8/11/2016	2180	209.93	6.44	14.72	31.564	09:25.1	15.21
8/12/2016	2180	196.70	6.50	15.07	31.569	09:28.9	14.91
8/13/2016	2180	183.48	6.55	15.42	31.575	09:32.7	14.61
8/14/2016	2180	170.26	6.60	15.77	31.580	09:36.4	14.30
8/15/2016	2180	157.04	6.65	16.11	31.586	09:40.2	13.99
8/16/2016	2180	143.82	6.70	16.44	31.592	09:43.9	13.67
8/17/2016	2180	130.60	6.74	16.77	31.598	09:47.7	13.36
8/18/2016	2180	117.38	6.79	17.10	31.604	09:51.4	13.03
8/19/2016	2180	104.16	6.83	17.42	31.610	09:55.1	12.71
8/20/2016	2180	90.94	6.87	17.74	31.616	09:58.8	12.38
8/21/2016	2180	77.72	6.91	18.05	31.622	10:2.5	12.05
8/22/2016	2180	64.50	6.94	18.36	31.629	10:6.2	11.71
8/23/2016	2180	51.28	6.98	18.66	31.635	10:9.9	11.37
8/24/2016	2180	38.07	7.01	18.96	31.642	10:13.6	11.03
8/25/2016	2180	24.85	7.04	19.25	31.649	10:17.2	10.69
8/26/2016	2180	11.64	7.07	19.54	31.656	10:20.9	10.34
8/27/2016	2180	358.42	7.09	19.82	31.663	10:24.5	9.99
8/28/2016	2181	345.21	7.12	20.09	31.670	10:28.2	9.64
8/29/2016	2181	331.99	7.14	20.36	31.677	10:31.8	9.29
8/30/2016	2181	318.84	7.16	20.63	31.684	10:35.5	8.93
8/31/2016	2181	305.63	7.18	20.89	31.692	10:39.1	8.57
9/1/2016	2181	292.42	7.20	21.14	31.699	10:42.7	8.21
9/2/2016	2181	279.21	7.21	21.39	31.707	10:46.4	7.84
9/3/2016	2181	265.99	7.22	21.64	31.714	10:50.0	7.48
9/4/2016	2181	252.78	7.23	21.87	31.722	10:53.6	7.11
9/5/2016	2181	239.57	7.24	22.11	31.730	10:57.2	6.74
9/6/2016	2181	226.36	7.24	22.33	31.738	11:0.8	6.36
9/7/2016	2181	213.16	7.25	22.55	31.746	11:4.4	5.99
9/8/2016	2181	199.95	7.25	22.77	31.754	11:8.0	5.61
9/9/2016	2181	186.74	7.25	22.97	31.762	11:11.6	5.24
9/10/2016	2181	173.53	7.25	23.18	31.770	11:15.2	4.86
9/11/2016	2181	160.33	7.24	23.37	31.778	11:18.8	4.48
9/12/2016	2181	147.12	7.24	23.57	31.787	11:22.4	4.10
9/13/2016	2181	133.91	7.23	23.75	31.795	11:26.0	3.72

(continued)

(continued)

Calendar Date	Rotation Number	Heliographic Lo	Bo	Po	Diameter (arcmin)	RA (HH:MM)	Dec (deg)
9/14/2016	2181	120.71	7.22	23.93	31.804	11:29.6	3.33
9/15/2016	2181	107.50	7.20	24.10	31.812	11:33.2	2.95
9/16/2016	2181	94.30	7.19	24.27	31.821	11:36.8	2.56
9/17/2016	2181	81.09	7.17	24.43	31.829	11:40.4	2.17
9/18/2016	2181	67.89	7.15	24.58	31.838	11:43.9	1.79
9/19/2016	2181	54.75	7.13	24.73	31.847	11:47.5	1.40
9/20/2016	2181	41.55	7.10	24.87	31.855	11:51.1	1.01
9/21/2016	2181	28.35	7.08	25.00	31.864	11:54.7	0.62
9/22/2016	2181	15.14	7.05	25.13	31.873	11:58.3	0.23
9/23/2016	2181	1.94	7.02	25.25	31.882	12:1.9	−0.16
9/24/2016	2182	348.74	6.99	25.37	31.891	12:5.5	−0.55
9/25/2016	2182	335.54	6.96	25.48	31.900	12:9.1	−0.94
9/26/2016	2182	322.34	6.92	25.58	31.909	12:12.7	−1.33
9/27/2016	2182	309.14	6.88	25.67	31.918	12:16.3	−1.72
9/28/2016	2182	295.94	6.84	25.76	31.927	12:19.9	−2.11
9/29/2016	2182	282.74	6.80	25.84	31.936	12:23.5	−2.50
9/30/2016	2182	269.55	6.75	25.92	31.945	12:27.1	−2.88
10/1/2016	2182	256.35	6.71	25.98	31.954	12:30.8	−3.27
10/2/2016	2182	243.15	6.66	26.04	31.964	12:34.4	−3.66
10/3/2016	2182	229.95	6.61	26.10	31.973	12:38.0	−4.05
10/4/2016	2182	216.76	6.56	26.14	31.982	12:41.7	−4.43
10/5/2016	2182	203.56	6.50	26.18	31.991	12:45.3	−4.82
10/6/2016	2182	190.36	6.45	26.21	32.000	12:48.9	−5.20
10/7/2016	2182	177.17	6.39	26.24	32.010	12:52.6	−5.59
10/8/2016	2182	163.97	6.33	26.26	32.019	12:56.3	−5.97
10/9/2016	2182	150.78	6.27	26.27	32.028	12:59.9	−6.35
10/10/2016	2182	137.65	6.20	26.27	32.037	13:3.6	−6.73
10/11/2016	2182	124.45	6.14	26.27	32.047	13:7.3	−7.11
10/12/2016	2182	111.26	6.07	26.25	32.056	13:11.0	−7.48
10/13/2016	2182	98.06	6.00	26.24	32.065	13:14.7	−7.86
10/14/2016	2182	84.87	5.93	26.21	32.074	13:18.4	−8.23
10/15/2016	2182	71.68	5.85	26.17	32.083	13:22.1	−8.60
10/16/2016	2182	58.48	5.78	26.13	32.092	13:25.8	−8.97
10/17/2016	2182	45.29	5.70	26.08	32.101	13:29.6	−9.34
10/18/2016	2182	32.10	5.62	26.03	32.110	13:33.3	−9.70
10/19/2016	2182	18.91	5.54	25.96	32.120	13:37.1	−10.06
10/20/2016	2182	5.71	5.46	25.89	32.129	13:40.9	−10.42
10/21/2016	2183	352.52	5.38	25.81	32.138	13:44.6	−10.78

(continued)

(continued)

Calendar	Rotation	Heliographic			Diameter	RA	Dec
Date	Number	*Lo*	*Bo*	*Po*	(arcmin)	(HH:MM)	(deg)
10/22/2016	2183	339.33	5.29	25.72	32.146	13:48.4	−11.13
10/23/2016	2183	326.14	5.21	25.63	32.155	13:52.2	−11.49
10/24/2016	2183	312.95	5.12	25.53	32.164	13:56.1	−11.84
10/25/2016	2183	299.76	5.03	25.41	32.173	13:59.9	−12.18
10/26/2016	2183	286.57	4.93	25.30	32.182	14:3.7	−12.52
10/27/2016	2183	273.38	4.84	25.17	32.191	14:7.6	−12.86
10/28/2016	2183	260.19	4.75	25.03	32.199	14:11.4	−13.20
10/29/2016	2183	247.00	4.65	24.89	32.208	14:15.3	−13.53
10/30/2016	2183	233.87	4.55	24.74	32.216	14:19.2	−13.86
10/31/2016	2183	220.68	4.45	24.58	32.225	14:23.1	−14.19
11/1/2016	2183	207.50	4.35	24.42	32.233	14:27.0	−14.51
11/2/2016	2183	194.31	4.25	24.25	32.242	14:30.9	−14.82
11/3/2016	2183	181.12	4.15	24.06	32.250	14:34.9	−15.14
11/4/2016	2183	167.93	4.04	23.87	32.258	14:38.8	−15.45
11/5/2016	2183	154.74	3.94	23.68	32.266	14:42.8	−15.75
11/6/2016	2183	141.55	3.83	23.47	32.274	14:46.8	−16.05
11/7/2016	2183	128.37	3.72	23.26	32.282	14:50.8	−16.35
11/8/2016	2183	115.18	3.61	23.04	32.290	14:54.8	−16.64
11/9/2016	2183	101.99	3.50	22.81	32.298	14:58.8	−16.93
11/10/2016	2183	88.81	3.39	22.57	32.306	15:2.8	−17.21
11/11/2016	2183	75.62	3.28	22.33	32.313	15:6.9	−17.49
11/12/2016	2183	62.43	3.16	22.08	32.321	15:10.9	−17.76
11/13/2016	2183	49.25	3.05	21.82	32.328	15:15.0	−18.03
11/14/2016	2183	36.06	2.93	21.55	32.335	15:19.1	−18.29
11/15/2016	2183	22.88	2.82	21.28	32.343	15:23.2	−18.55
11/16/2016	2183	9.69	2.70	21.00	32.350	15:27.3	−18.80
11/17/2016	2184	356.50	2.58	20.71	32.357	15:31.5	−19.04
11/18/2016	2184	343.32	2.46	20.41	32.364	15:35.6	−19.28
11/19/2016	2184	330.20	2.34	20.11	32.370	15:39.8	−19.52
11/20/2016	2184	317.01	2.22	19.80	32.377	15:44.0	−19.75
11/21/2016	2184	303.83	2.10	19.48	32.384	15:48.2	−19.97
11/22/2016	2184	290.64	1.98	19.16	32.390	15:52.4	−20.18
11/23/2016	2184	277.46	1.85	18.82	32.396	15:56.6	−20.39
11/24/2016	2184	264.28	1.73	18.48	32.403	16:0.8	−20.60
11/25/2016	2184	251.09	1.60	18.14	32.409	16:5.0	−20.80
11/26/2016	2184	237.91	1.48	17.79	32.415	16:9.3	−20.99
11/27/2016	2184	224.73	1.35	17.43	32.421	16:13.6	−21.17
11/28/2016	2184	211.54	1.23	17.06	32.426	16:17.8	−21.35

(continued)

(continued)

Calendar	Rotation	Heliographic			Diameter	RA	Dec
Date	Number	*Lo*	*Bo*	*Po*	(arcmin)	(HH:MM)	(deg)
11/29/2016	2184	198.36	1.10	16.69	32.432	16:22.1	−21.52
11/30/2016	2184	185.18	0.98	16.31	32.437	16:26.4	−21.68
12/1/2016	2184	172.00	0.85	15.93	32.443	16:30.7	−21.84
12/2/2016	2184	158.81	0.72	15.54	32.448	16:35.1	−21.99
12/3/2016	2184	145.63	0.59	15.15	32.453	16:39.4	−22.13
12/4/2016	2184	132.45	0.47	14.74	32.458	16:43.7	−22.27
12/5/2016	2184	119.27	0.34	14.34	32.462	16:48.1	−22.40
12/6/2016	2184	106.09	0.21	13.93	32.467	16:52.4	−22.52
12/7/2016	2184	92.91	0.08	13.51	32.471	16:56.8	−22.64
12/8/2016	2184	79.73	−0.05	13.09	32.476	17:1.2	−22.74
12/9/2016	2184	66.61	−0.17	12.66	32.480	17:5.6	−22.84
12/10/2016	2184	53.43	−0.30	12.23	32.484	17:10.0	−22.93
12/11/2016	2184	40.25	−0.43	11.79	32.488	17:14.4	−23.02
12/12/2016	2184	27.07	−0.56	11.35	32.491	17:18.8	−23.10
12/13/2016	2184	13.89	−0.69	10.91	32.495	17:23.2	−23.17
12/14/2016	2184	0.71	−0.81	10.46	32.498	17:27.6	−23.23
12/15/2016	2185	347.54	−0.94	10.01	32.501	17:32.0	−23.28
12/16/2016	2185	334.36	−1.07	9.56	32.504	17:36.4	−23.33
12/17/2016	2185	321.18	−1.19	9.10	32.507	17:40.9	−23.37
12/18/2016	2185	308.00	−1.32	8.63	32.510	17:45.3	−23.40
12/19/2016	2185	294.82	−1.45	8.17	32.513	17:49.7	−23.42
12/20/2016	2185	281.65	−1.57	7.70	32.515	17:54.2	−23.43
12/21/2016	2185	268.47	−1.70	7.23	32.517	17:58.6	−23.44
12/22/2016	2185	255.29	−1.82	6.76	32.519	18:3.0	−23.44
12/23/2016	2185	242.12	−1.95	6.29	32.521	18:7.5	−23.43
12/24/2016	2185	228.94	−2.07	5.81	32.523	18:11.9	−23.42
12/25/2016	2185	215.76	−2.19	5.33	32.525	18:16.4	−23.39
12/26/2016	2185	202.59	−2.31	4.85	32.526	18:20.8	−23.36
12/27/2016	2185	189.41	−2.44	4.37	32.527	18:25.	2-23.32
12/28/2016	2185	176.24	−2.56	3.89	32.528	18:29.7	−23.27
12/29/2016	2185	163.12	−2.68	3.41	32.529	18:34.1	−23.22
12/30/2016	2185	149.95	−2.80	2.92	32.530	18:38.5	−23.15
12/31/2016	2185	136.77	−2.91	2.44	32.531	18:42.9	−23.08

Index

Jamey L. Jenkins, *Observing the Sun: A Pocket Field Guide*,
Astronomer's Pocket Field Guide, DOI 10.1007/978-1-4614-8015-0,

Made in United States
Orlando, FL
21 April 2024

46031378R00146